# tredition®

tredition was established in 2006 by Sandra Latusseck and Soenke Schulz. Based in Hamburg, Germany, tredition offers publishing solutions to authors and publishing houses, combined with worldwide distribution of printed and digital book content. tredition is uniquely positioned to enable authors and publishing houses to create books on their own terms and without conventional manufacturing risks.

For more information please visit: www.tredition.com

## TREDITION CLASSICS

This book is part of the TREDITION CLASSICS series. The creators of this series are united by passion for literature and driven by the intention of making all public domain books available in printed format again - worldwide. Most TREDITION CLASSICS titles have been out of print and off the bookstore shelves for decades. At tredition we believe that a great book never goes out of style and that its value is eternal. Several mostly non-profit literature projects provide content to tredition. To support their good work, tredition donates a portion of the proceeds from each sold copy. As a reader of a TREDITION CLASSICS book, you support our mission to save many of the amazing works of world literature from oblivion. See all available books at www.tredition.com.

 Project Gutenberg

The content for this book has been graciously provided by Project Gutenberg. Project Gutenberg is a non-profit organization founded by Michael Hart in 1971 at the University of Illinois. The mission of Project Gutenberg is simple: To encourage the creation and distribution of eBooks. Project Gutenberg is the first and largest collection of public domain eBooks.

# Catalogue of Economic Plants in the Collection of the U. S. Department of Agriculture

William Saunders

# Imprint

This book is part of TREDITION CLASSICS

Author: William Saunders
Cover design: Buchgut, Berlin – Germany

Publisher: tredition GmbH, Hamburg - Germany
ISBN: 978-3-8472-1431-1

www.tredition.com
www.tredition.de

Copyright:
The content of this book is sourced from the public domain.

The intention of the TREDITION CLASSICS series is to make world literature in the public domain available in printed format. Literary enthusiasts and organizations, such as Project Gutenberg, worldwide have scanned and digitally edited the original texts. tredition has subsequently formatted and redesigned the content into a modern reading layout. Therefore, we cannot guarantee the exact reproduction of the original format of a particular historic edition. Please also note that no modifications have been made to the spelling, therefore it may differ from the orthography used today.

U. S. DEPARTMENT OF AGRICULTURE. CATALOGUE OF

# ECONOMIC PLANTS

IN THE COLLECTION OF THE
U. S. DEPARTMENT OF AGRICULTURE.

By WILLIAM SAUNDERS.

PUBLISHED BY AUTHORITY OF THE SECRETARY OF AGRICULTURE.

WASHINGTON:
GOVERNMENT PRINTING OFFICE.
1891.

[3]

# CATALOGUE OF ECONOMIC PLANTS IN THE COLLECTION OF THE U. S. DEPARTMENT OF AGRICULTURE.

U. S. Department of Agriculture,

*Washington, D. C., June 5, 1891.*

Sir: I have duly prepared by your direction a descriptive list of the more important economic plants at present contained in the collection of the Department, in such a form as will, in my opinion, most satisfactorily meet the wants of the numerous visitors and others interested in the work performed by the Department in this direction, and I beg to submit the same herewith for publication.

William Saunders,

*Superintendent of Gardens and Grounds.*

Hon. J. M. Rusk,

*Secretary of Agriculture.*

## DESCRIPTIVE CATALOGUE OF PLANTS.

1. Abelmoschus moschatus. — This plant is a native of Bengal. Its seeds were formerly mixed with hair powder, and are still used to perfume pomatum. The Arabs mix them with their coffee berries. In the West Indies the bruised seeds, steeped in rum, are used, both externally and internally, as a cure for snake bites.

2. Abrus precatorius. — Wild liquorice. This twining, leguminous plant is a native of the East, but is now found in the West Indies and other tropical regions. It is chiefly remarkable for its small oval seeds, which are of a brilliant scarlet color, with a black scar at the place where they are attached to the pods. These seeds are much used for necklaces and other ornamental purposes, and are employed in India as a standard of weight, under the name of Rati. The weight of the famous Kohinoor diamond is known to have been ascertained in this way. The roots afford liquorice, which is extracted in the same manner as that from the true Spanish liquorice plant, the *Glycyrrhiza glabra*. Recently the claim was made that the weather

could be foretold by certain movements of the leaves of this plant, but experimental tests have proved its fallacy.

3. Abutilon indicum.—This plant furnishes fiber fit for the manufacture of ropes. Its leaves contain a large quantity of mucilage.

4. Abutilon venosum.—This malvaceous plant is common in collections, as are others of the genus. They are mostly fiber-producing species. The flowers of *A. esculentum* are used as a vegetable in Brazil.

5. Acacia brasiliensis.—This plant furnishes the Brazil wood, which yields a red or crimson dye, and is used for dyeing silks. The best quality is that received from Pernambuco.

6. Acacia catechu.—The drug known as catechu is principally prepared from this tree, the wood of which is boiled down, and the decoction subsequently evaporated so as to form an extract much used as an astringent. The acacias are very numerous, and yield many useful products. Gum arabic is produced by several species, as *A. vera*, *A. arabica*, *A. adansonii*, *A. verek*, and others. It is obtained by spontaneous exudation from the trunk and branches, or by incisions made in the bark, from whence it flows in a liquid state, but [4] soon hardens by exposure to the air. The largest quantity of the gum comes from Barbary. Gum senegal is produced by *A. vera*. By some it is thought that the timber of *A. arabica* is identical with the Shittim tree, or wood of the Bible. From the flowers of *A. farnesiana* a choice and delicious perfume is obtained, the chief ingredient in many valued "balm of a thousand flowers." The pods of *A. concinna* are used in India as a soap for washing; the leaves are used for culinary purposes, and have a peculiarly agreeable acid taste. The seeds of some species are used, when cooked, as articles of food. From the seeds of *A. niopo* the Guahibo Indians prepare a snuff, by roasting the seeds and pounding them in a wooden platter. Its effects are to produce a kind of intoxication and invigorate the spirits. The bark of several species is extensively used for tanning, and the timber, being tough and elastic, is valuable for the manufacture of machinery and other purposes where great strength and durability are requisite.

7. Acacia dealbata.—The silver wattle tree of Australia. The bark is used for tanning purposes. It is hardy South.

8. Acacia homolophylla.—This tree furnishes the scented myall wood, a very hard and heavy wood, of an agreeable odor, resembling that of violets. Fancy boxes for the toilet are manufactured of it.

9. Acacia melanoxylon.—The wood of this tree is called mayall wood in New South Wales. It is also called violet wood, on account of the strong odor it has of that favorite flower; hence it is in great repute for making small dressing cases, etc.

10. Acacia mollissima.—The black wattle tree of Australia, which furnishes a good tanning principle. These trees were first called wattles from being used by the early settlers for forming a network or wattling of the supple twigs as a substitute for laths in plastering houses.

11. Acrocomia sclerocarpa.—This palm grows all over South America. It is known as the great macaw-tree. A sweetish-tasted oil, called Mucaja oil, is extracted from the fruit and is used for making toilet soaps.

12. Adansonia digitata.—The baobab tree, a native of Africa. It has been called the tree of a thousand years, and Humboldt speaks of it as "the oldest organic monument of our planet." Adanson, who traveled in Senegal in 1794, made a calculation to show that one of these trees, 30 feet in diameter, must be 5,150 years old. The bark of the baobab furnishes a fiber which is made into ropes and also manufactured into cloth. The fiber is so strong as to give rise to a common saying in Bengal, "as secure as an elephant bound with baobab rope." The pulp of the fruit is slightly acid, and the juice expressed from it is valued as a specific in putrid and pestilential fevers. The ashes of the fruit and bark, boiled in rancid palm oil, make a fine soap.

13. Adenanthera pavonina.—A tree that furnishes red sandal wood. A dye is obtained simply by rubbing the wood against a wet stone, which is used by the Brahmins for marking their foreheads after religious bathing. The seeds are used by Indian jewelers as weights, each seed weighing uniformly four grains. They are known as Circassian beans. Pounded and mixed with borax, they form an adhesive substance. They are sometimes used as food. The plant belongs to the *Leguminosæ*.

14. Adhatoda vasica. — This plant is extolled for its charcoal in the manufacture of powder. The flowers, leaves, roots, and especially the fruit, are considered antispasmodic, and are administered in India in asthma and intermittent fevers.

15. Ægle marmelos. — This plant belongs to the orange family, and its fruit is known in India as Bhel fruit. It is like an orange; the thick rind of the unripe fruit possesses astringent properties, and, when ripe, has an exquisite flavor and perfume. The fruit and other parts of the plant are used for medicinal purposes, and a yellow dye is prepared from the skin of the fruits.

16. Agave americana. — This plant is commonly known as American aloe, but it is not a member of that family, as it claims kindred with the *Amaryllis* tribe of plants. It grows naturally in a wide range of climate, from the plains of South America to elevations of 10,000 feet. It furnishes a variety of products. The plants form impenetrable fences; the leaves furnish fibers of various qualities, from the fine thread known as pita-thread, which is used for twine, to the coarse fibers used for ropes and cables. Humboldt describes a bridge of upward of 130 feet span over the Chimbo in Quito, of which the main ropes (4 inches in diameter) were made of this fiber. It is also used for making paper. The juice, when the watery part is evaporated, forms a good [5] soap (as detergent as castile), and will mix and form a lather with salt water as well as with fresh. The sap from the heart leaves is formed into pulque. This sap is sour, but has sufficient sugar and mucilage for fermentation. This vinous beverage has a filthy odor, but those who can overcome the aversion to this fetid smell indulge largely in the liquor. A very intoxicating brandy is made from it. Razor strops are made from the leaves; they are also used for cleaning and scouring pewter.

17. Agave rigida. — The sisal hemp, introduced into Florida many years ago, for the sake of its fiber, but its cultivation has not been prosecuted to a commercial success. Like many other of the best vegetable fibers found in leaves, it contains a gummy substance, which prevents the easy separation of the fiber from the pulp.

18. Aleurites triloba. — The candleberry tree, much cultivated in tropical countries for the sake of its nuts. The nuts or kernels, when dried and stuck on a reed, are used by the Polynesians as a substi-

tute for candles and as an article of food; they are said to taste like walnuts. When pressed, they yield largely of pure palatable oil, as a drying oil for paint, and known as artists' oil. The cake, after the oil has been expressed, is a favorite food for cattle. The root of the tree affords a brown dye, which is used to dye cloths.

19. Algarobia glandulosa.—The mezquite tree, of Texas, occasionally reaching a height of 25 to 30 feet. It yields a very hard, durable wood, and affords a large quantity of gum resembling gum arabic, and answering every purpose of that gum.

20. Allamanda cathartica.—This plant belongs to the family of *Apocynaceæ*, which contains many poisonous species. It is often cultivated for the beauty of its flowers; the leaves are considered a valuable cathartic, in moderate doses, especially in the cure of painter's colic; in large doses they are violently emetic. It is a native of South America.

21. Aloe socotrina.—Bitter aloe, a plant of the lily family, which furnishes the finest aloes. The bitter, resinous juice is stored up in greenish vessels, lying beneath the skin of the leaf, so that when the leaves are cut transversely, the juice exudes, and is gradually evaporated to a firm consistence. The inferior kinds of aloes are prepared by pressing the leaves, when the resinous juice becomes mixed with the mucilaginous fluid from the central part of the leaves, and thus it is proportionately deteriorated. Sometimes the leaves are cut and boiled, and the decoction evaporated to a proper consistence. This drug is imported in chests, in skins of animals, and sometimes in large calabash-gourds, and although the taste is peculiarly bitter and disagreeable, the perfume of the finer sorts is aromatic, and by no means offensive. It is common in tropical countries.

22. Alsophila australis.—This beautiful tree-fern attains a height of stem of 25 to 30 feet, with fronds spreading out into a crest 26 feet in diameter. These plants are among the most beautiful of all vegetable productions, and in their gigantic forms indicate, in a meager degree, the extraordinary beauty of the vegetation on the globe previous to the formation of the coal measures.

23. Alstonia scholaris.—The Pali-mara, or devil tree, of Bombay. The plant attains a height of 80 or 90 feet; the bark is powerfully

bitter, and is used in India in medicine. It is of the family of *Apocynaceæ*.

24. Amomum melegueta.—Malaguetta pepper, or grains of paradise; belonging to the ginger family, *Zingiberaceæ*. The seeds of this and other species are imported from Guinea; they have a very warm and camphor-like taste, and are used to give a fictitious strength to adulterated liquors, but are not considered particularly injurious to health. The seeds are aromatic and stimulating, and form, with other seeds of similar plants, what are known as cardamoms.

25. Amyris balsamifera.—This plant yields the wood called Lignum Rhodium. It also furnishes a gum resin analogous to Elemi, and supposed to yield Indian Bdellium.

26. Anacardium occidentale.—The cashew nut tree, cultivated in the West Indies and other tropical countries. The stem furnishes a milky juice, which becomes hard and black when dry, and is used as a varnish. It also secretes a gum, like gum arabic. The nut or fruit contains a black, acrid, caustic oil, injurious to the lips and tongue of those who attempt to crack the nut with their teeth; it becomes innocuous and wholesome when roasted, but this process must be carefully conducted, the acridity of the fumes producing severe inflammation of the face if approached too near. [6]

27. Ananassa sativa.—The well-known pineapple, the fruit of which was described three hundred years ago, by Jean de Léry, a Huguenot priest, as being of such excellence that the gods might luxuriate upon it, and that it should only be gathered by the hand of a Venus. It is supposed to be a native of Brazil, and to have been carried from thence to the West, and afterwards to the East Indies. It first became known to Europeans in Peru. It is universally acknowledged to be one of the most delicious fruits in the world. Like all other fruits that have been a long time under cultivation, there are numerous varieties that vary greatly, both in quality and appearance. The leaves yield a fine fiber, which is used in the manufacture of pina cloth; this cloth is very delicate, soft, and transparent, and is made into shawls, scarfs, handkerchiefs, and dresses.

28. Andira inermis.—This is a native of Senegambia. Its bark is anthelmintic, but requires care in its administration, being powerfully narcotic. It has a sweetish taste, but a disagreeable smell, and is

generally given in the form of a decoction, which is made by boiling an ounce of the dried bark in a quart of water until it assumes the color of Madeira wine. Three or four grains of the powdered bark acts as a powerful purgative. The bark is known as bastard cabbage bark, or worm bark. It is almost obsolete in medicine.

29. Andropogon muricatus. — The Khus-Khus, or Vetiver grass of India. The fibrous roots yield a most peculiar but pleasing perfume. In India the leaves are manufactured into awnings, blinds, and sunshades; but principally for screens, used in hot weather for doors and windows, which, when wetted, diffuse a peculiar and refreshing perfume, while cooling the air.

30. Andropogon schœnanthus. — The sweet-scented lemon grass, a native of Malabar. An essential oil is distilled from the leaves, which is used in perfumery. It is a favorite herb with the Asiatics, both for medicinal and culinary purposes. Tea from the dried leaves is a favorite beverage of some persons.

31. Anona cherimolia. — The Cherimoyer of Peru, where it is extensively cultivated for its fruits, which are highly esteemed by the inhabitants, but not so highly valued by those accustomed to the fruits of temperate climates. The fruit, when ripe, is of a pale greenish-yellow color, tinged with purple, weighing from 3 to 4 pounds; the skin thin; the flesh sweet, and about the consistence of a custard; hence often called custard apple.

32. Anona muricata. — The sour-sop, a native of the West Indies, which produces a fruit of considerable size, often weighing over 2 pounds. The pulp is white and has an acrid flavor, which is not disagreeable.

33. Anona reticulata. — The common custard apple of the West Indies. It has a yellowish pulp and is not so highly esteemed as an article of food as some others of the species. It bears the name of Condissa in Brazil. The Anonas are grown to some extent throughout southern Florida.

34. Anona squamosa. — The sweet-sop, a native of the Malay Islands, where it is grown for its fruits. These are ovate in shape, with a thick rind, which incloses a luscious pulp. The seeds contain an

acrid principle, and, being reduced to powder, form an ingredient for the destruction of insects.

35. Antiaris innoxia.—The upas tree. Most exaggerated statements respecting this plant have passed into history. Its poisonous influence was said to be so great as not only to destroy all animal life but even plants could not live within 10 miles of it. The plant has no such virulent properties as the above, but, as it inhabits low valleys in Java where carbonic acid gas escapes from the crevices in volcanic rocks which frequently proves fatal to animals, the tree was blamed wrongly. It is, however, possessed of poisonous juice, which, when dry and mixed with other ingredients, forms a venomous poison for arrows, and severe effects have been felt by those who have climbed upon the branches for the purpose of gathering the flowers.

36. Antiaris saccidora.—The sack tree; so called from the fibrous bark being used as sacks. For this purpose young trees of about a foot in diameter are selected and cut into junks of the same length as the sack required. The outer bark is then removed and the inner bark loosened by pounding, so that it can be separated by turning it inside out. Sometimes a small piece of the wood is left to form the bottom of the sack. The fruit exudes a milky, viscid juice, which hardens into the consistency of beeswax, but becomes black and shining.

37. Antidesma bunias.—An East India plant which produces small, intensely black fruit about the size of a currant, used in making preserves. The bark [7] furnishes a good fiber, which is utilized in the manufacture of ropes. A decoction of the leaves is a reputed cure for snake bites. The whole plant is very bitter.

38. Aralia papyrifera.—The Chinese rice paper plant. The stems are filled with pith of very fine texture and white as snow, from which is derived the article known as rice paper, much used in preparing artificial flowers.

39. Araucaria bidwillii.—The Bunya-Bunya of Australia, which forms a large tree, reaching from 150 to 200 feet in height. The cones are very large, and contain one hundred to one hundred and fifty seeds, which are highly prized by the aborigines as food. They are best when roasted in the shell, cracked between two stones and

eaten while hot. In flavor they resemble roasted chestnuts. During the season of the ripening of these seeds the natives grow sleek and fat. That part of the country where these trees most abound is called the Bunya-Bunya country.

40. Araucaria brasiliensis.—The Brazilian Araucaria, which grows at great elevations. The seeds of this tree are commonly sold in the markets of Rio Janeiro as an article of food. The resinous matter which exudes from the trunk is employed in the manufacture of candles.

41. Araucaria cunninghamii.—The Morton Bay pine. This Australian tree forms a very straight trunk, and yields a timber of much commercial importance in Sidney and other ports. It is chiefly used for house building and some of the heavier articles of furniture.

42. Araucaria excelsa.—This very elegant evergreen is a native of Norfolk Island. Few plants can compare with it in beauty and regularity of growth. The wood is of no particular value, although used for building purposes in Norfolk Island.

43. Ardisia crenata.—A native of China. The bark has tonic and astringent properties, and is used in fevers and for external application in the cure of ulcers, etc.

44. Areca catechu.—This palm is cultivated in all the warmer parts of Asia for its seed. This is known under the name of betel nut, and is about the size of a nutmeg. The chewing of these nuts is a common practice of hundreds of thousands of people. The nut is cut into small pieces, mixed with a small quantity of lime, and rolled up in leaves of the betel pepper. The pellet is chewed, and is hot and acrid, but possesses aromatic and astringent properties. It tinges the saliva red and stains the teeth. The practice is considered beneficial rather than otherwise, just as chewing tobacco-leaves, drinking alcohol, and eating chicken-salad are considered healthful practices in some portions of the globe. A kind of catechu is obtained by boiling down the seeds to the consistence of an extract, but the chief supply of this drug is Acacia catechu.

45. Argania sideroxylon.—This is the argan tree of Morocco. It is remarkable for its low-spreading mode of growth. Trees have been measured only 16 feet in height, while the circumference of the

branches was 220 feet. The fruit is much eaten and relished by cattle. The wood is hard and so heavy as to sink in water. A valuable oil is extracted from the seeds.

46. Aristolochia grandiflora.—The pelican flower. This plant belongs to a family famed for the curious construction of their flowers, as well as for their medical qualities. In tropical America various species receive the name of "Guaco," which is a term given to plants that are used in the cure of snake bites. Even some of our native species, such as *A. serpentaria,* is known as snake-root, and is said to be esteemed for curing the bite of the rattlesnake. It is stated that the Egyptian jugglers use some of these plants to stupefy the snakes before they handle them. *A. bracteata* and *A. indica* are used for similar purposes in India. It is said that the juice of the root of *A. anguicida,* if introduced into the mouth of a serpent, so stupefies it that it may be handled with impunity. The Indians, after having "guaconized" themselves, that is, having taken Guaco, handle the most venomous snakes without injury.

47. Artanthe elongata.—A plant of the pepper family, which furnishes one of the articles known by the Peruvians as Matico, and which is used by them for the same purposes as cubebs; but its chief value is as a styptic, an effect probably produced by its rough under surface, acting mechanically like lint. It has been employed internally to check hemorrhages, but with doubtful [8] effect. Its aromatic bitter stimulant properties are like those of cubebs, and depend on a volatile oil, a dark-green resin, and a peculiar bitter principle called *maticin.*

48. Artocarpus incisa.—This is the breadfruit tree of the South Sea Islands, where its introduction gave occasion for the historical incidents arising from the mutiny of the "Bounty." The round fruits contain a white pulp, of the consistence of new bread. It is roasted before being eaten, but has little flavor. The tree furnishes a viscid juice containing caoutchouc, which is used as glue for calking canoes. In the South Sea Islands the breadfruit constitutes the principal article of diet; it is prepared by baking in an oven heated by hot stones.

49. Artocarpus integrifolia.—The jack of the Indian Archipelago, cultivated for its fruit, which is a favorite article among the natives,

as also are the roasted seeds. The wood is much used, and resembles mahogany. Bird-lime is made from the juice.

50. Astrocaryum vulgare.—Every part of this South American palm is covered with sharp spines. It is cultivated to some extent by the Indians of Brazil for the sake of its young leaves, which furnish a strong fiber for making bowstrings, fishing nets, etc. The finer threads are knitted into hammocks, which are of great strength. It is known as Tucum thread. The pulp of the fruit furnishes an oil. In Guiana it is called the Aoura palm.

51. Attalea cohune.—This palm furnishes Cahoun nuts, from which is extracted cohune oil, used as a burning oil, for which purpose it is superior to cocoanut oil. Piassaba fiber is furnished by this and *A. funifera*, the seeds of which are known as Coquilla nuts; these nuts are 3 or 4 inches long, oval, of a rich brown color, and very hard; they are much used by turners for making the handles of doors, umbrellas, etc. The fiber derived from the decaying of the cellular matter at the base of the leaf-stalks is much used in Brazil for making ropes. It is largely used in England and other places for making coarse brooms, chiefly used in cleaning streets.

52. Averrhoa bilimbi.—This is called the blimbing, and is cultivated to some extent in the East Indies. The fruit is oblong, obtuse-angled, somewhat resembling a short, thick cucumber, with a thin, smooth, green rind, filled with a pleasant, acid juice.

53. Averrhoa carambola.—The caramba of Ceylon and Bengal. The fruit of this tree is about the size of a large orange, and, when ripe, is of a rich yellow color, with a very decided and agreeable fragrance. The pulp contains a large portion of acid, and is generally used as a pickle or preserve. In Java it is used both in the ripe and unripe state in pies; a sirup is also made of the juice, and a conserve of the flowers. These preparations are highly valued as remedies in fevers and bilious disorders.

54. Bactris major.—The Marajah palm, of Brazil, which grows upon the banks of the Amazon River. It has a succulent, rather acid fruit, from which a vinous beverage is prepared. *B. minor* has a stem about 14 feet high and about an inch in diameter. These stems are used for walking canes, and are sometimes called Tobago canes.

55. Balsamocarpon brevifolium.—This shrub is the algarrobo of the Chilians. It belongs to the pea family. Its pods are short and thick, and when unripe contain about 80 per cent of tannic acid; the ripe pods become transformed into a cracked resinous substance, when their tanning value is much impaired; this resinous matter is astringent, and is used for dyeing black and for making ink.

56. Balsamodendron myrrha.—A native of Arabia Felix, producing a gum resin, sometimes called Opobalsamum, which was considered by the ancients as a panacea for almost all the ills that flesh is heir to. *B. mukul* yields a resin of this name, and is considered identical with the Bdellium of Dioscorides and of the Scriptures. The resin has cordial and stimulating properties, and is burnt as an incense. In ancient times it was used as an embalming ingredient.

57. Bambusa arundinacea.—The bamboo cane, a gigantic grass, cultivated in many tropical and semitropical countries. The Chinese use it in one way or other for nearly everything they require. Almost every article of furniture in their houses, including mats, screens, chairs, tables, bedsteads, and bedding, is made of bamboo. The masts, sails, and rigging of their ships consist chiefly of bamboo. A fiber has been obtained from the stem suitable for mixing [9] with wool, cotton, and silk; it is said to be very soft and to take dyes easily. They have treatises and volumes on its culture, showing the best soil and the seasons for planting and transplanting this useful production.

58. Bauhinia vahlii.—The Maloo-climber of India, where the gigantic shrubby stems often attain a height of 300 feet, running over the tops of the tallest trees, and twisting so tightly around their stems as to kill them. The exceedingly tough fibrous bark of this plant is used in India for making ropes and in the construction of suspension bridges. The seeds form an article of food; they are eaten raw, and resemble cashew nuts in flavor.

59. Beaucarnea recurvifolia.—This Mexican plant is remarkable for the large bulbiform swelling at the base of the stem. It is a plant of much elegance and beauty, resembling a drooping fountain.

60. Bergera koenigii.—The curry-leaf tree of India. The fragrant, aromatic leaves are used to flavor curries. The leaves, root, and bark

are used medicinally. The wood is hard and durable, and from the seeds a clear, transparent oil, called Simbolee oil, is extracted.

61. Berrya ammonilla.—This furnishes the Trincomalee wood of the Philippine Islands and Ceylon, and is largely used for making oil casks and for building boats, for which it is well adapted, being light and strong.

62. Bertholletia excelsa.—This furnishes the well known Brazil nuts, or cream nuts of commerce. The tree is a native of South America and attains a height of 100 to 150 feet. The fruit is nearly round and contains from eighteen to twenty-four seeds, which are so beautifully packed in the shell that when once removed it is found impossible to replace them. A bland oil is pressed from the seeds, which is used by artists, and at Para the fibrous bark of the tree is used for calking ships, as a substitute for oakum.

63. Bignonia echinata.—A native of Mexico, where it is sometimes called Mariposa butterfly. The branches are said to be used in the adulteration of sarsaparilla. *B. chica*, a native of Venezuela, furnishes a red pigment, obtained by macerating the leaves in water, which is used by the natives for painting their bodies. The long flexible stems of *B. kerere* furnish the natives of French Guiana with a substitute for ropes. *B. alliacea* is termed the Garlic shrub, because of the powerful odor of garlic emitted from its leaves and branches when bruised. These plants all have showy flowers, and the genus is represented with us by such beautiful flowers as are produced by *B. radicans* and *B. capreolata*.

64. Bixa orellana.—Arnotta plant. This plant is a native of South America, but has been introduced and cultivated both in the West and East Indies. It bears bunches of pink-colored flowers, which are followed by oblong bristled pods. The seeds are thinly coated with red, waxy pulp, which is separated by stirring them in water until it is detached, when it is strained off and evaporated to the consistence of putty, when it is made up into rolls; in this condition it is known as flag or roll arnotta, but when thoroughly dried it is made into cakes and sold as cake arnotta. It is much used by the South American Caribs and other tribes of Indians for painting their bodies, paint being almost their only article of clothing. As a commercial article it is mainly used as a coloring for cheese, butter, and

inferior chocolates, to all of which it gives the required tinge without imparting any unpleasant flavor or unwholesome quality. It is also used in imparting rich orange and gold-colored tints to various kinds of varnishes.

65. Blighia sapida.—The akee fruit of Guinea. The fruit is about 3 inches long by 2 inches wide; the seeds are surrounded by a spongy substance, which is eaten. It has a subacid, agreeable taste. A small quantity of semisolid fatty oil is obtained from the seeds by pressure.

66. Bœhmeria nivea.—A plant of the nettle family, which yields the fiber known as Chinese grass. The beautiful fabric called grasscloth, which rivals the best French cambric in softness and fineness of texture, is manufactured from the fiber of this plant. The fiber is also variously known in commerce as rheea, ramie, and in China as Tchow-ma. It is a plant of the easiest culture, and has been introduced into the Southern States, where it grows freely. When once machinery is perfected so as to enable its being cheaply prepared for the manufacturer, a great demand will arise for this fiber.

67. Boldoa fragrans.—A Chilian plant which yields small edible fruits; these, as well as all parts of the plant, are very aromatic. The bark is used for tanning, and the wood is highly esteemed for making charcoal. An alkaloid [10] called *boldine*, extracted from the plant, has reputed medicinal value, and a drug called Boldu is similarly produced.

68. Borassus flabelliformis.—The Palmyra palm. The parts of this tree are applied to such a multitude of purposes that a poem in the Tamil language, although enumerating eight hundred uses, does not exhaust the catalogue. In old trees the wood becomes hard and is very durable. The leaves are from 8 to 10 feet long, and are used for thatching houses, making various mattings, bags, etc. They also supply the Hindoo with paper, upon which he writes with a stylus. A most important product called toddy or palm wine is obtained from the flower spikes, which yield a great quantity of juice for four or five months. Palm-toddy is intoxicating, and when distilled yields strong arrack. Very good vinegar is also obtained from it, and large quantities of jaggery or palm sugar are manufactured from the toddy. The fruits are large and have a thick coating of fibrous pulp,

which is cooked and eaten or made into jelly. The young palm plants are cultivated for the market, as cabbages are with us, and eaten, either when fresh or after being dried in the sun.

69. Boswellia thurifera.—This Coromandel tree furnishes the resin known as olibanum, which is supposed to have been the frankincense of the ancients. It is sometimes used in medicine as an astringent and stimulant, and is employed, because of its grateful perfume, as an incense in churches.

70. Bromelia karatas.—The Corawa fiber, or silk-grass of Guiana, is obtained from this plant, which is very strong, and much used for bowstrings, fishing lines, nets, and ropes.

71. Bromelia pinguin.—This is very common as a hedge or fence plant in the West Indies. The leaves, when beaten with a blunt mallet and macerated in water, produce fibers from which beautiful fabrics are manufactured. The fruit yields a cooling juice much used in fevers.

72. Brosimum alicastrum.—The bread-nut tree of Jamaica. The nuts or seeds produced by this tree are said to form an agreeable and nutritious article of food. When cooked they taste like hazelnuts. The young branches and shoots are greedily eaten by horses and cattle, and the wood resembles mahogany, and is used for making furniture.

73. Brosimum galactodendron.—The cow tree of South America, which yields a milk of as good quality as that from the cow. It forms large forests on the mountains near the town of Cariaco and elsewhere along the seacoast of Venezuela, reaching to a considerable height. In South America the cow tree is called Palo de Vaca, or Arbol de Leche. Its milk, which is obtained by making incisions in the trunk, so closely resembles the milk of the cow, both in appearance and quality, that it is commonly used as an article of food by the inhabitants of the places where the tree is abundant. Unlike many other vegetable milks, it is perfectly wholesome, and very nourishing, possessing an agreeable taste, and a pleasant balsamic odor, its only unpleasant quality being a slight amount of stickiness. The chemical analysis of this milk has shown it to possess a composition closely resembling some animal substances; and, like animal milk, it quickly forms a cheesy scum, and after a few days' exposure

to the atmosphere, turns sour and putrefies. It contains upwards of 30 per cent of a resinous substance called *galactine*.

74. Brya ebenus.—Jamaica or West India ebony tree. This is not the plant that yields the true ebony-wood of commerce. Jamaica ebony is of a greenish-brown color, very hard, and so heavy that it sinks in water. It takes a good polish, and is used by turners for the manufacture of numerous kinds of small wares.

75. Byrsonima spicata.—A Brazilian plant, furnishing an astringent bark used for tanning, and also containing a red coloring matter employed in dyeing. The berries are used in medicine, and a decoction of the roots is used for ulcers.

76. Cæsalpinia bonduc.—A tropical plant, bearing the seeds known as nicker nuts, or bonduc nuts. These are often strung together for necklaces. The kernels have a very bitter taste, and the oil obtained from them is used medicinally.

77. Cæsalpinia pulcherrima.—This beautiful flowering leguminous plant is a native of the East Indies, but is cultivated in all the tropics. In Jamaica it is called the "Barbados flower." The wood is sought after for charcoal, and a decoction of the leaves and flowers is used in fevers. [11]

78. Cæsalpinia sappan.—The brownish-red wood of this Indian tree furnishes the Sappan wood of commerce, from which dyers obtain a red color, principally used for dyeing cotton goods. Its root also affords an orange-yellow dye.

79. Calamus rotang.—This is one of the palms that furnish the canes or rattans used for chair bottoms, sides of pony-carriages, and similar purposes. It is a climbing palm and grows to an immense length; specimens 300 feet long have been exhibited, climbing over and amongst the branches of trees, supporting themselves by means of the hooked spines attached to the leaf stalks. *C. rudentum* and *C. viminalis* furnish flexible canes. In their native countries they are used for a variety of manufacturing purposes, also for ropes and cables used by junks and other coasting vessels. In the Himalayas they are used in the formation of suspension bridges across rivers and deep ravines. *C. scipionum* furnishes the well-known Malacca canes used for walking sticks. They are naturally of a rich brown

color. The clouded and mottled appearance which some of these present is said to be imparted to them by smoking and steaming.

80. Callistemon salignus.—A medium-sized tree from Australia; one of the many so-called tea trees of that country. The wood, which is very hard, is known as stone wood and has been used for wood engraving. Layers of the bark readily peel off; hence it also receives the name of paper-bark plant.

81. Callitris quadrivalvis.—This coniferous plant is a native of Barbary. It yields a hard, durable, and fragrant timber, and is much employed in the erection of mosques, etc., by the Africans of the North. The resin that exudes from the tree is used in varnish under the name of gum-sandarach. In powder it forms a principal ingredient of the article known as pounce.

82. Calophyllum calaba.—This is called calaba tree in the West Indies, and an oil, fit for burning, is expressed from the seeds. In the West Indies these seeds are called Santa Maria nuts.

83. Calotropis gigantea.—The inner bark of this plant yields a valuable fiber, capable of bearing a greater strain than hemp. All parts of it abound in a very acrid milky juice, which hardens into a substance resembling gutta-percha; but in its fresh state it is a valuable remedy in cutaneous diseases. The bark of the root also possesses similar medical qualities; and its tincture yields *mudarine*, a substance that has the property of gelatinizing when heated, and returning to the fluid state when cool. Paper has been made from the silky down of the seeds.

84. Camellia japonica.—A well-known green-house plant, cultivated for its large double flowers. The seeds furnish an oil of an agreeable odor, which is used for many domestic purposes.

85. Camphora officinarum.—This tree belongs to the *Lauraceæ*. Camphor is prepared from the wood by boiling chopped branches in water, when, after some time, the camphor becomes deposited and is purified by sublimation. It is mainly produced in the island of Formosa. The wood of the tree is highly prized for manufacturing entomological cabinets. As the plant grows well over a large area in the more Southern States, it is expected that the preparation of its products will become a profitable industry.

86. Canella alba.—This is a native of the West Indies, and furnishes a pale olive-colored bark with an aromatic odor, and is used as a tonic. It is used by the natives as a spice. It furnishes the true canella bark of commerce, also known as white-wood bark.

87. Capparis spinosa.—The caper plant, a native of the South of Europe and of the Mediterranean regions. The commercial product consists of the flower-buds, and sometimes the unripe fruits, pickled in vinegar. The wood and bark possess acrid qualities which will act as a blister when applied to the skin.

88. Carapa guianensis.—A meliaceous plant, native of tropical America, where it grows to a height of 60 to 80 feet. The bark of this tree possesses febrifugal properties and is also used for tanning. By pressure, the seeds yield a liquid oil called carap-oil or crab-oil, suitable for burning in lamps.

89. Carica papaya.—This is the South American papaw tree, but is cultivated in most tropical countries. It is also known as the melon-apple. The fruit is of a dingy orange-color, of an oblong form, about 8 to 10 inches long, by 3 or 4 inches broad. It is said that the juice of the tree, or an infusion of the leaves and fruit, has the property of rendering tough fiber quite tender. Animals fed upon the fruit and leaves will have very tender and juicy flesh. [12]

90. Carludovica palmata.—A pandanaceous plant from Panama and southward. Panama hats are made from the leaves of this plant. The leaves are cut when young, and the stiff parallel veins removed, after which they are slit into shreds, but not separated at the stalk end, and immersed in boiling water for a short time, then bleached in the sun.

91. Caryocar nuciferum.—On the river banks of Guiana this grows to a large-sized tree. It yields the butter-nuts, or souari-nuts of commerce. These are of a flattened kidney shape, with a hard woody shell of a reddish-brown color, and covered with wart-like protuberances. The nuts are pleasant to eat, and yield, by expression, an oil called Piquia oil, which possesses the flavor of the fruit.

92. Caryophyllus aromaticus.—This myrtaceous plant produces the well-known spice called cloves. It forms a beautiful evergreen, rising from 20 to 30 feet in height. The cloves of commerce are the

unexpanded flower-buds; they are collected by beating the tree with rods, when the buds, from the jointed character of their stalks, readily fall, and are received on sheets spread on purpose; they are then dried in the sun. All parts of the plant are aromatic, from the presence of a volatile oil. The oil is sometimes used in toothache and as a carminative in medicine.

93. Caryota urens.—This fine palm is a native of Ceylon, and is also found in other parts of India, where it supplies the native population with various important articles. Large quantities of toddy, or palm-wine, are prepared from the juice, which, when boiled, yields very good palm sugar or jaggery, and also excellent sugar candy. Sago is also prepared from the central or pithy part of the trunk, and forms a large portion of the food of the natives. The fiber from the leaf stalk is of great strength; it is known as Kittool fiber, and is used for making ropes, brushes, brooms, etc. A woolly kind of scurf, scraped off the leaf stalks, is used for calking boats, and the stem furnishes a small quantity of wood.

94. Casimiroa edulis.—A Mexican plant, belonging to the orange family, with a fruit about the size of an ordinary orange, which has an agreeable taste, but is not considered to be wholesome. The seeds are poisonous; the bark is bitter, and is sometimes used medicinally.

95. Cassia acutifolia.—The cassias belong to the leguminous family. The leaflets of this and some other species produce the well-known drug called senna. That known as Alexandria senna is produced by the above. East Indian senna is produced by *C. elongata*. Aleppo senna is obtained from *C. obovata*. The native species, *C. marylandica*, possesses similar properties. The seeds of *C. absus*, a native of Egypt, are bitter, aromatic, and mucilaginous, and are used as a remedy for ophthalmia. *C. fistula* is called the Pudding-Pipe tree, and furnishes the cassia pods of commerce. The seeds of *C. occidentalis*, when roasted, are used as a substitute for coffee in the Mauritius and in the interior of Africa.

96. Castilloa elastica.—This is a Mexican tree, which yields a milky juice, forming caoutchouc, but is not collected for commerce except in a limited way.

97. Casuarina quadrivalvis.—This Tasmanian tree produces a very hard wood of a reddish color, often called Beef wood. It is

marked with dark stripes, and is much used in some places for picture frames and cabinetwork. This belongs to a curious family of trees having no leaves, but looking like a gigantic specimen of Horse-tail grass, a weed to be seen in wet places.

98. Catha edulis.—This plant is a native of Arabia, where it attains the height of 7 to 10 feet. Its leaves are used by the Arabs in preparing a beverage like tea or coffee. The twigs, with leaves attached, in bundles of fifty, and in pieces from 12 to 15 inches in length, form a very considerable article of commerce, its use in Arabia corresponding to that of the Paraguay tea in South America and the Chinese tea in Europe. The effects produced by a decoction of the leaves of Cafta, as they are termed, are described as similar to those produced by strong green tea, only more pleasing and agreeable. The Arab soldiers chew the leaves when on sentry duty to keep them from feeling drowsy. Its use is of great antiquity, preceding that of coffee. Its stimulating effects induced some Arabs to class it with intoxicating substances, the use of which is forbidden by the Koran, but a synod of learned Mussulmans decreed that, as it did not impair the health or impede the observance of religious duties, but only increased hilarity and good humor, it was lawful to use it.

99. Cecropia peltata.—The South American trumpet tree, so called because its hollow branches are used for musical instruments. The Waupe Indians form [13] a kind of drum by removing the pith or center of the branches. The inner bark of the young branches yields a very tough fiber, which is made into ropes. The milky juice of the stem hardens into caoutchouc.

100. Cedrela odorata.—This forms a large tree in the West India Islands, and is hollowed out for canoes; the wood is of a brown color and has a fragrant odor, and is sometimes imported under the name of Jamaica cedar.

101. Cephælis ipecacuanha.—This Brazilian plant produces the true ipecacuanha, and belongs to the *Cinchonaceæ*. The root is the part used in medicine, it is knotty, contorted, and annulated, and of a grayish-brown color, and its emetic properties are due to a chemical principle called *emetin*.

102. Ceratonia siliqua.—The carob bean. This leguminous plant is a native of the countries bordering on the Mediterranean. The seed

pods contain a quantity of mucilaginous and saccharine matter, and are used as food for cattle. Besides the name of carob beans, these pods are known as locust pods, or St. John's bread, from a supposition that they formed the food of St. John in the wilderness. It is now generally admitted that the locusts of St. John were the insects so called, and which are still used as an article of food in some of the Eastern countries. There is more reason for the belief that the husks mentioned in the parable of the prodigal son were these pods. The seeds were at one time used by singers, who imagined that they softened and cleared the voice.

103. Cerbera thevetia.—The name is intended to imply that the plant is as dangerous as Cerberus. The plant has a milky, poisonous juice. The bark is purgative; the unripe fruit is used by the natives of Travancore to destroy dogs, as its action causes their teeth to loosen and fall out.

104. Cereus gigantea.—The suwarrow of the Mexicans, a native of the hot, arid, and almost desert regions of New Mexico, found growing in rocky places, in valleys, and on mountain sides, often springing out of mere crevices in hard rocks, and imparting a singular aspect to the scenery of the country, its tall stems often reaching 40 feet in height, with upright branches looking like telegraph posts for signaling from point to point of the rocky mountains. The fruits are about 2 or 3 inches long, of a green color and oval form; when ripe they burst into three or four pieces, which curve back so as to resemble a flower. Inside they contain numerous little black seeds, imbedded in a crimson-colored pulp, which the Indians make into a preserve. They also eat the ripe fruit as an article of food.

105. Cereus macdonaldiæ.—A night-blooming cereus, and one of the most beautiful. The flowers when fully expanded are over a foot in diameter, having numerous radiating red and bright orange sepals and delicately white petals. It is a native of the Honduras.

106. Ceroxylon andicola.—The wax palm of New Grenada, first described by Humboldt and Bonpland, who found it on elevated mountains, extending as high as the lower limit of perpetual snow. Its tall trunk is covered with a thin coating of a whitish waxy substance, giving it a marbled appearance. The waxy substance forms an article of commerce, and is obtained by scraping the trunk. It

consists of two parts of resin and one wax, and, when mixed with one third of tallow, it makes very good candles. The stem is used for building purposes, and the leaves for thatching roofs.

107. Chamædorea elegans.—This belongs to a genus of palms native of South America. The plant is of tall, slender growth; the stems are used for walking canes, and the young, unexpanded flower spikes are used as a vegetable.

108. Chamærops fortunei.—This palm is a native of the north of China, and is nearly hardy here. In China, the coarse brown fibers obtained from the leaves are used for making hats and also garments called So-e, worn in wet weather.

109. Chamærops humilis.—This is the only European species of palm, and does not extend farther north than Nice. The leaves are commonly used in the south of Europe for making hats, brooms, baskets, etc. From the leaf fiber a material resembling horse hair is prepared, and the Arabs mix it with camel's hair for their tent covers.

110. Chavica betel.—This plant is found all over the East Indies, where its leaf is largely used by Indian natives as a masticatory. Its consumption is immense, [14] and has been said to equal that of tobacco by Western peoples. It is prepared for chewing by inclosing in the leaves a slice of the areca nut, and a small portion of lime. It is thought to act as a stimulant to the digestive organs, but causes giddiness and other unpleasant symptoms to those not accustomed to its use.

111. Chiococca racemosa.—This plant is found in many warm countries, such as in southern Florida. It is called cahinca in Brazil, where a preparation of the bark of the root is employed as a remedy for snake bites. Almost every locality where snakes exist has its local remedies for poisonous bites, but they rarely prove to be efficient when truthfully and fairly tested.

112. Chloranthus officinalis.—The roots of this plant are an aromatic stimulant, much used as medicine in the Island of Java; also, when mixed with anise, it has proved valuable in malignant smallpox.

113. Chloroxylon swietenia.—The satinwood tree of tropical countries. It is principally used for making the backs of clothes and hair brushes, and for articles of turnery-ware; the finest mottled pieces are cut into veneers and used for cabinet-making.

114. Chrysobalanus icaco.—The cocoa plum of the West Indies. The fruits are about the size of a plum, and are of various colors, white, yellow, red, or purple. The pulp is sweet, a little austere, but not disagreeable. The fruits are preserved and exported from Cuba and other West India Islands. The kernels yield a fixed oil, and an emulsion made with them is used medicinally.

115. Chrysophyllum cainito.—The fruit of this plant is known in the West Indies as the star apple, the interior of which, when cut across, shows ten cells, and as many seeds disposed regularly round the center, giving a star-like appearance, as stars are generally represented in the most reliable almanacs. It receives its botanic name from the golden silky color on the under side of the leaves.

116. Cicca disticha.—This Indian plant is cultivated in many parts under the name of Otaheite gooseberry. The fruits resemble those of a green gooseberry. They have an acid flavor; are used for preserving or pickling, and eaten either in a raw state or cooked in various ways.

117. Cinchona calisaya.—The yellow bark of Bolivia. This is one of the so-called Peruvian Bark trees. The discovery of the medicinal value of this bark is a matter of fable and conjecture. The name cinchona is derived from that of the wife of a viceroy of Peru, who is said to have taken the drug from South America to Europe in 1639. Afterwards the Jesuits used it; hence it is sometimes called Jesuit's bark. It was brought most particularly into notice when Louis XIV of France purchased of Sir R. Talbor, an Englishman, his heretofore secret remedy for intermittent fever, and made it public.

There are various barks in commerce classified under the head of Peruvian barks. Their great value depends upon the presence of certain alkaloid substances called quinine, cinchonine, and quinidine, which exist in the bark in combination with tannic and other acids. Quinine is the most useful of these alkaloids, and this is found in greatest quantities in Calisaya bark. The gray bark of Huanuco is derived from *Cinchona micrantha*, which is characterized by its yield of cinchonine, and the Loxa or Loja barks are furnished in part by *Cinchona officinalis*, and are especially rich in quinidine. There is some uncertainty about the trees that produce the various kinds of bark. These trees grow in the forests of Bolivia and Peru, at various elevations on the mountains, but chiefly in sheltered mountain valleys, and all of them at a considerable distance below the frost or snow line. They are destroyed by the slightest frost. Plants of various species have been distributed from time to time, in localities which seemed most favorable to their growth, but all reports from these distributions have, so far, been discouraging.

118. Cinnamomum cassia.—This furnishes cassia bark, which is much like cinnamon, but thicker, coarser, stronger, less delicate in flavor, and cheaper; hence it is often used to adulterate cinnamon. The unexpanded flower buds are sold as cassia buds, possessing properties similar to those of the bark. It is grown in southern China, Java, and tropical countries generally.

119. Cinnamomum zeylanicum.—A tree belonging to *Lauraceæ*, which furnishes the best cinnamon. It is prepared by stripping the bark from the branches, when it rolls up into quills, the smaller of which are introduced into the [15] larger, and then dried in the sun. Cinnamon is much used as a condiment for its pleasant flavor, and its astringent properties are of medicinal value. It is cultivated largely in Ceylon. The cinnamon tree is too tender to become of commercial importance in the United States. Isolated plants may be found in southern Florida, at least it is so stated, but the area suited to its growth must be very limited.

120. Cissampelos pareira.—The velvet plant of tropical countries. The root furnishes the *Pareira brava* of druggists, which is used in medicine.

121. Citrus aurantium.—The orange, generally supposed to be a native of the north of India. It was introduced into Arabia during the ninth century. It was unknown in Europe in the eleventh century. Oranges were cultivated at Seville towards the end of the twelfth century, and at Palermo in the thirteenth. In the fourteenth century they were plentiful in several parts of Italy. There are many varieties of the orange in cultivation. The blood red, or Malta, is much esteemed; the fruit is round, reddish-yellow outside and the pulp

irregularly mottled with crimson. The Mandarin or Tangerine orange has a thin rind which separates easily from the pulp, and is very sweet and rich. The St. Michael's orange is one of the most productive and delicious varieties, with a thin rind and very sweet pulp. The Seville or bitter orange is used for the manufacture of bitter tincture and candied orange-peel. The Bergamot orange has peculiarly fragrant flowers and fruit, from each of which an essence of a delicious quality is extracted.

122. Citrus decumana.—The shaddock, which has the largest fruit of the family. It is a native of China and Japan, where it is known as sweet ball. The pulp is acid or subacid, and in some varieties nearly sweet. From the thickness of the skin the fruit will keep a considerable time without injury.

123. Citrus japonica.—This is the Kum-quat of the Chinese. It forms a small tree, or rather a large bush, and bears fruit about the size of a large cherry. There are two forms, one bearing round fruits, the other long, oval fruits. This fruit has a sweet rind and an agreeably acid pulp, and is usually eaten whole without being peeled. It forms an excellent preserve, with sugar, and is largely used in this form.

124. Citrus limetta.—The lime, which is used for the same purposes as the lemon, and by some preferred, the juice being considered more wholesome and the acid more agreeable. There are several varieties, some of them being sweet and quite insipid.

125. Citrus limonum.—The lemon; this plant is found growing naturally in that part of India which is beyond the Ganges. It was unknown to the ancient Greeks and Romans. It is supposed to have been brought to Italy by the Crusaders. Arabian writers of the twelfth century notice the lemon as being cultivated in Egypt and other places. The varieties of the lemon are very numerous and valued for their agreeable acid juice and essential oil. They keep for a considerable time, especially if steeped for a short period in salt water.

126. Citrus medica.—The citron, found wild in the forests of northern India. The Jews cultivated the citron at the time they were under subjection to the Romans, and used the fruit in the Feast of the Tabernacles. There is no proof of their having known the fruit in

the time of Moses, but it is supposed that they found it at Babylon, and brought it into Palestine. The citron is cultivated in China and Cochin-China. It is easily naturalized and the seeds are rapidly spread. In its wild state it grows erect; the branches are spiny, the flowers purple on the outside and white on the inside. The fruit furnishes the essential oil of citron and the essential oil of cedra. There are several varieties; the fingered citron is a curious fruit, and the Madras citron is very long and narrow; the skin is covered with protuberances.

127. Clusia rosea.—A tropical plant which yields abundantly of a tenacious resin from its stem, which is used for the same purpose as pitch. It is first of a green color, but when exposed to the air it assumes a brown or reddish tint. The Caribs use it for painting the bottoms of their boats.

128. Coccoloba uvifera.—Known in the West Indies as the seaside grape, from the peculiarity of the perianth, which becomes pulpy and of a violet color and surrounds the ripe fruit. The pulpy perianth has an agreeable acid flavor. An astringent extract is prepared from the plant which is used in medicine.

129. Cocos nucifera.—The cocoanut palm. This palm is cultivated throughout the tropics so extensively that its native country is not known. One reason [16] of its extensive dissemination is that it grows so close to the sea that the ripe fruits are washed away by the waves and afterwards cast upon far-distant shores, where they soon vegetate. It is in this way that the coral islands of the Indian Ocean have become covered with these palms. Every part of this tree is put to some useful purpose. The outside rind or husk of the fruit yields the fiber from which the well-known cocoa matting is manufactured. Cordage, clothes, brushes, brooms, and hats are made from this fiber, and, when curled and dyed, it is used for stuffing mattresses and cushions. An oil is produced by pressing the white kernel of the nut which is used for cooking when fresh, and by pressure affords stearin, which is made into candles, the liquid being used for lamps. The kernel is of great importance as an article of food, and the milk affords an agreeable beverage. While young it yields a delicious substance resembling blanc-mange. The leaves are used for thatching, for making mats, baskets, hats, etc.; combs are

made from the hard footstalk; the heart of the tree is used as we use cabbages. The brown fibrous net work from the base of the leaves is used as sieves, and also made into garments. The wood is used for building and for furniture. The flowers are used medicinally as an astringent and the roots as a febrifuge.

130. Cocos plumosus.—A Brazilian species, highly ornamental in its long, arching leaves, and producing quantities of orange-colored nuts, in size about as large as a chestnut, inclosed in an edible pulp.

131. Coffea arabica.—The coffee plant, which belongs to the *Cinchonaceæ* and is a native of Abyssinia, but is now cultivated in many tropical regions. It can not be successfully cultivated in a climate where the temperature, at any season of the year, falls below 55 degrees, although it will exist where the temperature all but falls short of freezing, but a low fall of temperature greatly retards the ripening of the fruit. Ripe fruits are often gathered from plants in the extreme south of Florida. The beans or seeds are roasted before use, and by this process they gain nearly one half in bulk and lose about a fifth in weight. Heat also changes their essential qualities, causing the development of the volatile oil and peculiar acid to which the aroma and flavor are due. The berries contain theine; so also do the leaves, and in some countries the latter are preferred.

132. Coffea liberica.—The Liberian coffee, cultivated in Africa, of which country it is a native. This plant is of larger and stronger growth than the Arabian coffee plant and the fruit is larger. This species is of recent introduction to commerce, and although it was reported as being more prolific than the ordinary coffee plant, the statement has not been borne out in Brazil and Mexico, where it has been tested. It is also more tender than the older known species.

133. Cola acuminata.—An African tree, which has been introduced into the West Indies and Brazil for the sake of its seeds, which are known as Cola, or Kola, or Goora nuts, and extensively used as a sort of condiment by the natives of Africa. A small piece of one of these seeds is chewed before each meal to promote digestion. It possesses properties similar to the leaves of coca and contains theine. These nuts have from time immemorial occupied a prominent place in the dietetic economy of native tribes in Africa, and the

demand for them has established a large commercial industry in the regions where they are obtained.

134. Colocasia esculenta.—This plant has been recommended for profitable culture in this country for its edible root-stock. It is cultivated in the Sandwich Islands under the name of Tara. The young leaves are cooked and eaten in the same manner as spinach or greens in Egypt. They are acrid, but lose their acridity when boiled, the water being changed. The roots are filled with starch, and have long been used as food in various semitropical countries.

135. Condaminea macrophylla.—This plant belongs to the cinchona family, and contains tonic properties. The Peruvian bark gatherers adulterate the true cinchona bark with this, but it may be detected by its white inner surface, its less powerful bitter taste, and a viscidity not possessed by the cinchonas.

136. Convolvulus scammonia.—This plant furnishes the scammony of the druggists.

137. Cookia punctata.—A small-growing tree from China, which produces a fruit known as the Wampee. This fruit is a globular berry, with five or fewer compartments filled with juice. It is much esteemed in China. [17]

138. Copaifera officinalis.—This tree yields balsam of copaiba, used in medicine. The balsam is collected by making incisions in the stem, when the liquor is said to pour out copiously; as it exudes it is thin and colorless, but immediately thickens and changes to a clear yellow. Like many other balsams, it is nearly allied to the turpentines; it has a moderately agreeable smell, and a bitter, biting taste of considerable duration. Distilled with water it yields a limpid essential oil.

139. Copernica cerifera.—The Carnuba, or wax palm of Brazil. It grows about 40 feet high, and has a trunk 6 or 8 inches thick, composed of very hard wood, which is commonly employed in Brazil for building and other purposes. The upper part of the young stem is soft, and yields a kind of sago, and the bitter fruits are eaten by the Indians. The young leaves are coated with wax, called Carnaub wax, which is detached by shaking them, and then melted and run into cakes; it is harder than beeswax, and has been used for making

candles. The leaves are used for thatch, and, when young, are eaten by cattle.

140. Coprosma robusta.—A cinchonaceous shrub. The leaves of this plant were formerly used in some of the religious ceremonies of the New Zealanders.

141. Cordia myxa.—This produces succulent, mucilaginous, and emollient fruits, which are eaten. These qualities, combined with a slight astringency, have led to their use as pectorals, known as Sebestens. The wood of this tree is said to have furnished the material used by the Egyptians in the construction of their mummy cases; it is also considered to be one of the best woods for kindling fire by friction.

142. Cordyline australis.—The Australian Ti, or cabbage tree, a palm-like plant of 15 to 20 feet in height. The whole plant is fibrous, and it has been suggested as good for a paper-making material. The juice of the roots and stem contains a small amount of sugar, and has been employed for procuring alcohol.

143. Corypha umbraculifera.—The Talipot palm, a native of Ceylon, producing gigantic fan-like leaves. These leaves have prickly stalks 6 or 7 feet long, and when fully expanded form a nearly complete circle of 13 feet in diameter. Large fans made of these leaves are carried before people of rank among the Cinghalese; they are also commonly used as umbrellas, and tents are made by neatly joining them together; they are also used as a substitute for paper, being written upon with a stylus. Some of the sacred books of the Cinghalese are composed of strips of them. The hard seeds are used by turners.

144. Couroupita guianensis.—The fruit of this tree is called, from its appearance, the cannon-ball fruit; its shell is used as a drinking vessel, and when fresh the pulp is of an agreeable flavor.

145. Cratæva gynandra.—This West Indian tree yields a small fruit which has a strong smell of garlic, hence it is called the garlic pear. The bark is bitter and used as a tonic.

146. Crescentia cujete.—The calabash tree of the West Indies, where it is valued for the sake of its fruits, which resemble pumpkins in appearance and occasionally reach a diameter of 18 inches.

Divested of their pulp, which is not edible, they serve various useful domestic purposes, for carrying water, and even as kettles for cooking. They are strong and light.

147. Croton balsamiferum.—This West Indian shrub is sometimes called seaside balsam or sage. A thick, yellowish, aromatic juice exudes from the extremities of the broken branches, or wherever the stem has been wounded. In Martinique a liquor called *Eau de Mantes* is distilled from this balsamic juice with spirits of wine. The young leaves and branches are used in warm baths, on account of their agreeable fragrance and reputed medicinal virtues.

148. Croton eleutheria.—This plant furnishes cascarilla bark, used as an aromatic bitter tonic, having no astringency. It has a fragrant smell when burnt, on which account it has been mixed with smoking tobacco.

149. Croton tiglium.—A plant of the family *Euphorbiaceæ*, from the Indian Archipelago, which produces the seeds from whence croton oil is extracted. It is a very powerful medicine, and even in pressing the seeds for the purpose of extracting the oil, the workmen are subject to irritation of the eyes and other casualties. [18]

150. Cubeba officinalis.—A native of Java, which furnishes the cubeb fruits of commerce. These fruits are like black pepper, but stalked, and have an acrid, hot, aromatic taste; frequently used medicinally.

151. Curcas purgans.—A tropical plant cultivated in many warm countries for the sake of its seeds, known as physic nuts. The juice of the plant, which is milky, acrid, and glutinous, produces an indelible brown stain on linen. The oil from the seeds is used for burning in lamps; and in paints. In China it is boiled with oxide of iron and used as a varnish. It is also used medicinally.

152. Curcuma longa.—A plant belonging to the *Zingiberaceæ*, the roots of which furnish turmeric. This powder is used in India as a mild aromatic, and for other medicinal purposes. It also enters into the composition of curry-powder, and a sort of arrowroot is made from the young tubers.

153. Curcuma zedoaria.—This plant furnishes zedoary tubers, much used in India as aromatic tonics.

154. Cyathea medullaris.—This beautiful tree fern is a native of Australia, where it attains a height of 25 to 30 feet, having fronds from 10 to 15 feet in length. It contains a pulpy substance in the center of the stem, of a starchy, mucilaginous nature, which is a common article of food with the natives. The trees have to be destroyed in order to obtain it.

155. Cybistax antisyphilitica.—A plant of the order of *Bignoniaceæ*, called Atunyangua in the Andes of Peru, where the inhabitants dye their cotton clothes by boiling them along with the leaves of this plant; the dye is a permanent blue. The bark of the young shoots is much employed in medicine.

156. Cycas revoluta.—The sago palm of gardens. The stem of the plants abounds in starch, which is highly esteemed in Japan. A gum exudes from the trunk of the old plant, which is employed medicinally by the natives of India.

157. Cycas circinalis.—A native of Malabar, where a kind of sago is prepared from the seeds, which are dried and powdered; medicinal properties are also attributed to the seeds.

158. Dacrydium franklinii.—Called Huon pine, because of its being found near the Huon River, in Tasmania. It belongs to the yew family. It furnishes valuable timber, very durable, and is used for ship and house building; some of the wood is very beautifully marked, and is used in furniture making and cabinetwork.

159. Dalbergia sissoo.—A tree of northern India, the timber of which is known as Sissum wood. This wood is strong, tenacious, and compact, much used for railway ties and for gun-carriages.

160. Damara australis.—A singular plant of the *Coniferæ* family, called the Kauri pine. It forms a tree 150 to 200 feet in height, and produces a hard, brittle resin-like copal, which is used in varnish.

161. Dasylirion acrotrichum.—A plant of the pineapple family, from Mexico. The leaves contain a fine fiber, which may be ultimately more extensively utilized than it is at present.

162. Desmodium gyrans.—An interesting plant of the pea family, called the moving plant, on account of the rotatory motion of the leaflets. These move in all conceivable ways, either steadily or by

jerks. Sometimes only one leaf or two on the plant will be affected; at other times a nearly simultaneous movement may be seen in all the leaves. These movements are most energetic when the thermometer marks about 80°. This motion is not due to any external or mechanical irritation.

163. Dialium acutifolium.—The velvet tamarind, so called, from the circumstance that its seed-pods are covered with a beautiful black velvet down. The seeds are surrounded by a farinaceous pulp of an agreeable acid taste.

164. Dialium indum.—The tamarind plum, which has a delicious pulp of slightly acid flavor.

165. Dicksonia antarctica.—The large fern tree of Australia. This plant attains the height of 30 or more feet, and its fronds or leaves spread horizontally some 20 to 25 feet. It is found in snowy regions, and would be perfectly hardy south. It is one of the finest objects of the vegetable kingdom when of sufficient size to show its true beauties.

166. Dieffenbachia seguina.—This has acquired the name of dumb cane, in consequence of its fleshy, cane-like stems, rendering speechless any person [19] who may happen to bite them, their acrid poison causing the tongue to swell to an immense size. An ointment for applying to dropsical swellings is prepared by boiling the juice in lard. Notwithstanding its acridity, a wholesome starch is prepared from the stem.

167. Dillenia speciosa.—An East Indian tree, bearing a fruit which is used in curries and for making jellies. Its slightly acid juice, sweetened with sugar, forms a cooling beverage. The wood is very tough, and is used for making gun-stocks.

168. Dion edule.—A Mexican plant, bearing large seeds containing a quantity of starch, which is separated and used as arrowroot.

169. Diospyros ebenum.—An East Indian tree which in part yields the black ebony wood of commerce, much used in fancy cabinetwork and turnery, door knobs, pianoforte keys, etc.

170. Diospyros kaki.—The Chinese date plum or persimmon. The fruits vary in size from that of a medium-sized apple to that of a

large pear; they also vary much in their flavor and consistency, some being firm, and others having a soft custard-like pulp, very sweet and luscious. The Chinese dry them in the sun and make them into sweetmeats; they are sometimes imported, and in appearance resemble large-sized preserved figs. These plants are being quite largely cultivated in some of the southern States, and the fruit is entering commerce.

171. Dipterix odorata.—This leguminous plant yields the fragrant seed known as Tonka bean, used in scenting snuff and for other purposes of perfumery. The odor resembles that of new-mown hay, and is due to the presence of *coumarine*. The tree is a native of Cayenne and grows 60 to 80 feet high.

172. Dorstenia contrayerva.—A plant from tropical America, the roots of which are used in medicine under the name of Contrayerva root.

173. Dracæna draco.—The Dragon's Blood tree of Teneriffe. This liliaceous plant attains a great age and enormous size. The resin obtained from this tree has been found in the sepulchral caves of the Cuanches, and hence it is supposed to have been used by them in embalming the dead. Trees of this species, at present in vigorous health, are supposed to be as old as the pyramids of Egypt.

174. Dracænopsis Australis.—Ti or cabbage tree of New Zealand. The whole of this plant is fibrous and has been used for paper making. The juice of the roots and stem contains a small amount of sugar and has been used for producing alcohol.

175. Drimys winteri.—This plant belongs to the magnolia family and furnishes the aromatic tonic known as Winter's bark. It is a native of Chili and the Strait of Magalhaens.

176. Dryobalanops aromatica.—A native of the Island of Sumatra. It furnishes a liquid called camphor oil and a crystalline solid known as Sumatra or Borneo camphor. Camphor oil is obtained from incisions in the tree, and has a fragrant, aromatic odor. It has been used for scenting soap. The solid camphor is found in cracks of the wood, and is obtained by cutting down the tree, dividing it into blocks and small pieces, from the interstices of which the camphor is extracted. It differs from the ordinary camphor in being more

brittle and not condensing on the sides of the bottle in which it is kept. It is much esteemed by the Chinese, who attribute many virtues to it. It has been long known and is mentioned by Marco Polo in the thirteenth century.

177. Duboisia hopwoodii.—The leaves of this Australian plant are chewed by the natives of Central Australia, just as the Peruvians and Chilians masticate the leaves of the *Erythroxylon coca*, to invigorate themselves during their long foot journeys through the country. They are known as Pitury leaves.

178. Durio zibethinus.—A common tree in the Malayan Islands, where its fruit forms a great part of the food of the natives. It is said to have a most delicious flavor combined with a most offensive odor, but when once the repugnance of the peculiar odor is overcome it becomes a general favorite. The unripe fruit is cooked and eaten, and the seeds roasted and used like chestnuts.

179. Elæis guineensis.—The African oil palm is a native of southwestern Africa, but has been introduced into other regions. It grows to a height of 20 to 30 [20] feet and bears dense heads of fruit. The oil is obtained by boiling the fruits in water and skimming off the oil as it rises to the surface. It is used in the manufacture of candles. In Africa it is eaten as butter by the natives.

180. Elæis melanococca.—A palm from tropical America which produces large quantities of oil.

181. Elæocarpus hinau.—A New Zealand tree, of the linden family. The bark affords an excellent permanent dye, varying from light brown to deep black. The fruits are surrounded by an edible pulp, and they are frequently pickled like olives.

182. Elettaria cardamomum.—This plant furnishes the fruits known as the Small or Malabar cardamoms of commerce. The seeds are used medicinally for their cordial aromatic properties, which depend upon the presence of a volatile oil. In India the fruits are chewed by the natives with their betel.

183. Emblica officinalis.—A plant belonging to *Euphorbiaceæ*, a native of India. In Borneo the bark and young shoots are used to dye cotton black, for which purpose they are boiled in alum. The fruits are made into sweetmeats, with sugar, or eaten raw, but they are

exceedingly acid; when ripe and dry, they are used in medicine, under the name of *Myrobalani emblici*. The natives of Travancore have a notion that the plant imparts a pleasant flavor to water, and therefore place branches of the tree in their wells, especially when the water is charged with an accumulation of impure vegetable matter.

184. Enckea unguiculata.—A plant of the family *Piperaceæ*, having an aromatic fruit like a berry, with a thick rind. The roots are used medicinally in Brazil.

185. Entada scandens.—This leguminous plant has remarkable pods, which often measure 6 or 8 feet in length. The seeds are about 2 inches across, and half an inch thick, and have a hard, woody, and beautifully polished shell, of a dark-brown or purplish color. These seeds are frequently converted into snuff-boxes and other articles, and in the Indian bazars they are used as weights.

186. Eriodendron anfractuosum.—The silk-cotton, or God tree of the West Indies. The fruit is a capsule, filled with a beautiful silky fiber, which is very elastic, but can not be woven, and is only used for stuffing cushions.

187. Erythrina caffra.—The Kaffir tree of South Africa. The wood is soft and so light as to be used for floating fishing nets. The scarlet seeds are employed for making necklaces. The Erythrinas, of which there are many species, are mostly remarkable for the brilliant scarlet of their flowers, and are known as Coral trees.

188. Erythrina umbrosa.—This is a favorite tree for growing in masses, for the purpose of sheltering cocoanut plantations, and inducing a proper degree of moisture in their neighborhood.

189. Erythroxylon coca.—The leaves of this plant, under the name of coca, are much used by the inhabitants of South America as a masticatory. It forms an article of commerce among the Indians, who carefully dry the leaves and use them daily. Their use, in moderation, acts as a stimulant to the nervous system and enables those who chew them to perform long journeys without any other food. The use of coca in Peru is a very ancient custom, said to have originated with the Incas. It is common throughout the greater part of Peru, Quito, New Granada; and on the banks of the Rio Negro it is

known as Spadic. A principle, called *cocaine*, has been extracted from the leaves, which is used in medicine.

190. Eucalyptus amygdalina.—The peppermint tree, a native of Tasmania. It produces a thin, transparent oil possessed of a pungent odor resembling oil of lemons, and tasting like camphor, which has great solvent properties. The genus *Eucalyptus* is extensive and valuable. The greater number form large trees, known in Australia as gum trees.

191. Eucalyptus gigantea.—This stringy bark gum furnishes a strong, durable timber, used for shipbuilding and other purposes. *E. robusta* contains large cavities in its stem, between the annual concentric circles of wood, filled with a red gum. Many of the species yield gums and astringent principles and also a species of manna. The timber of these trees has been pronounced to be unsurpassed for strength and durability by any other timber known. The leaves of these trees are placed vertically to the sun, a provision suited to a dry and sultry climate. [21]

192. Eucalyptus globulus.—The blue gum, a rapid-growing tree, attaining to a large size. Recently it has attracted attention and gained some repute in medicine as an antiperiodic. The leaves have also been applied to wounds with some success. It produces a strong camphor-smelling oil, which has a mint-like taste, not at all disagreeable.

193. Eugenia acris.—The wild clove or bayberry tree of the West Indies. In Jamaica it is sometimes called the black cinnamon. The refreshing perfume known as bay rum is prepared by distilling the leaves of this tree with rum. It is stated that the leaves of the allspice are also used in this preparation.

194. Eugenia jambosa.—A tropical plant, belonging to the myrtle family, which produces a pleasant rose-flavored fruit, known as the Roseapple, or Jamrosade.

195. Eugenia pimento.—The fruits of this West Indian tree are known in commerce as allspice; the berries have a peculiarly grateful odor and flavor, resembling a combination of cloves, nutmeg, and cinnamon; hence the name of allspice. The leaves when bruised emit a fine aromatic odor, and a delicate odoriferous oil is distilled

from them, which is said to be used as oil of cloves. The berries, bruised and distilled with water, yield the pimento oil of commerce.

196. Eugenia ugni.—This small-foliaged myrtaceous plant is a native of Chili. It bears a glossy black fruit, which has an agreeable flavor and perfume, and is highly esteemed in its native country. The plant is hardy in the Southern States.

197. Euphorbia canariensis.—This plant grows in abundance in the Canary Islands and Teneriffe, in dry, rocky districts, where little else can grow, and where it attains a height of 10 feet, with the branches spreading 15 or 20 feet. It is one of the kinds that furnish the drug known as *Euphorbium*. The milky juice exudes from incisions made in the branches, and is so acrid that it excoriates the hand when applied to it. As it hardens it falls down in small lumps, and those who collect it are obliged to tie cloths over their mouths and nostrils to exclude the small, dusty particles, as they produce incessant sneezing. As a medicine its action is violent, and it is now rarely employed. There are a vast number of species of *Euphorbia*, varying exceedingly in their general appearance, but all of them having a milky juice which contains active properties. Many of them can scarcely be distinguished from cactuses so far as relates to external appearances, but the milky exudation following a puncture determines their true character. *E. grandidens* is a tall-growing, branching species, and attains a height of 30 feet. The natives of India use the juice of *E. antiquorum*, when diluted, as a purgative. The juice of *E. heptagona* and other African species is employed to poison arrows; the juice of *E. cotinifolia* is used for the same purpose in Brazil. The roots of *E. gerardiana* and *E. pithyusa* are emetic, while *E. thymifolia* and *E. hypericifolia* possess astringent and aromatic properties. The poisonous principle which pervades these plants is more or less dissipated by heat. The juice of *E. cattimandoo* furnishes caoutchouc of a very good quality, which, however, becomes brittle, although soaking in hot water renders it again pliable. *E. phosphorea* derives the name from the fact of its sap emitting a phosphorescent light, on warm nights, in the Brazilian forests.

198. Euterpe edulis.—The assai palm of Para. It grows in swampy lands, and produces a small fruit thinly coated with clotted flesh of which the inhabitants of Para manufacture a beverage called assai.

The ripe fruits are soaked in warm water and kneaded until the fleshy pulp is detached. This, when strained, is of a thick, creamy consistence, and, when thickened with cassava farina and sweetened with sugar, forms a nutritious diet, and is the daily food of a large number of the people.

199. Euterpe montana.—The center portion of the upper part of the stem of this West Indian palm, including the leaf bud, is eaten either when cooked as a vegetable or pickled, but the tree must be destroyed in order to obtain it.

200. Excœcaria sebifera.—This Euphorbiaceous plant is the tallow tree of China. The fruits, are about half an inch in diameter, and each contains three seeds, thickly coated with a fatty substance which yields the tallow. This is obtained by first steaming the seeds, then bruising them to loosen the fat without breaking the seeds, which are removed by sifting. The fat is then made into flat circular cakes and pressed, when the pure tallow exudes [22] in a liquid state and soon hardens into a white, brittle mass. Candles made from this get soft in hot weather, which is prevented by coating them with insect wax. A liquid oil is obtained from the seeds by pressing. The tree yields a hard wood, used by the Chinese for printing blocks, and its leaves are used in dyeing black.

201. Exogonium purga.—This plant furnishes the true jalap-tubers of commerce. They owe their well-known purgative properties to their resinous ingredients. Various species of Ipomœa furnish a spurious kind of this drug, which is often put in the market as the genuine article.

202. Exostemma caribæum.—This West Indian plant has become naturalized in southern Florida. It belongs to the cinchona family and is known as Jamaica bark. It is also known as Quinquina Caraibe. The bark is reputed to be a good febrifuge, and also to be employed as an emetic. It is supposed to contain some peculiar principle, as the fracture displays an abundance of small crystals. The capsules, before they are ripe, are very bitter, and their juice causes a burning itching on the lips.

203. Feronia elephantum.—The wood apple or elephant apple tree of India, belonging to the family *Aurantiaceæ*. It forms a large tree in Ceylon, and yields a hard, heavy wood, of great strength. It

yields a gum, which is mixed with other gums and sold under the name of East Indian gum arabic. The fruit is about the size of an orange, and contains a pulpy flesh, which is edible, and a jelly is made from it, which is used in cases of dysentery. The leaves have an odor like that of anise, and the native India doctors employ them as a stomachic and carminative.

204. Fevillea cordifolia. — The sequa or cacoon antidote of Jamaica. It belongs to the cucumber family, and climbs to a great height up the trunks of trees. The seeds are employed as a remedy in a variety of diseases, and are considered an antidote against the effects of poison; they also contain a quantity of semisolid fatty oil, which is liberated by pressing and boiling them in water.

205. Ficus elastica. — This plant is known as the india-rubber tree. It is a native of the East Indies, and is the chief source of caoutchouc from that quarter of the globe, although other species of Ficus yield this gum, as well as several plants of other genera. It is a plant of rapid growth, and from the larger branches roots descend to the earth as in the case of the banyan tree.

206. Ficus indica. — The famous banyan tree of history. Specimens of this Indian fig are mentioned as being of immense size. One in Bengal spreads over a diameter of 370 feet. Another covered an area of 1,700 square yards. It is one of the sacred trees of the Hindoos. It was known to the ancients. Strabo describes it, and it is mentioned by Pliny. Milton also alludes to it as follows:

> Branching so broad along, that in the ground
> The bending twigs take root; and daughters grow
> About the mother tree; a pillared shade,
> High overarched, with echoing walks between.
> There oft the Indian herdsman, shunning heat,
> Shelters in cool; and tends his pasturing herds
> At loop-holes cut through thickest shade.

207. Ficus religiosa. — The pippul tree of the Hindoos, which they hold in such veneration that, if a person cuts or lops off any of the branches, he is looked upon with as great abhorrence as if he had

broken the leg of one of their equally sacred cows. The seeds are employed by Indian doctors in medicine.

208. Flacourtia sepiaria.—A bushy shrub, used in India for hedges. Its fruit has a pleasant, subacid flavor when perfectly ripe, but the unripe fruit is extremely astringent. The Indian doctors use a liniment made of the bark in cases of gout, and an infusion of it as a cure for snake bites.

209. Fourcroya cubense.—This plant is closely related to the agave, and, like many of that genus, furnishes a fine fiber, which is known in St. Domingo as Cabuya fiber. These plants are very magnificent when in flower, throwing up stems 20 to 30 feet in height, covered with many hundreds of yucca-like blossoms.

210. Franciscea uniflora.—A Brazilian plant called Mercurio vegetal; also known as Manaca. The roots, and to some extent the leaves, are used in medicine; the inner bark and all the herbaceous parts are nauseously bitter; it is regarded [23] as a purgative, emetic, and alexipharmic; in overdoses it is an acrid poison.

211. Fusanus acuminatus.—A small tree of the Cape of Good Hope and Australia. It bears a globular fruit of the size of a small peach, and is known in Australia as the native peach. It has an edible nut, called the Quandang nut, which is said to be as sweet and palatable as the almond.

212. Galipea officinalis.—This South American tree furnishes Angostura bark, which has important medical properties, some physicians in South America preferring it to cinchona in the treatment of fevers. Its use has been greatly retarded by bark of the deadly nux-vomica tree having been inadvertently sold for it. As this bark is sometimes used in bitters, a mistake, as above, might prove as fatal as cholera.

213. Garcinia mangostana.—This tree produces the tropical fruit called mangosteen, a beautiful fruit, having a thick, succulent rind, which contains an astringent juice, and exudes a gum similar to gamboge. The esculent interior contains a juicy pulp, of the whiteness and solubility of snow, and of a refreshing, delicate, delicious flavor. The bark of the tree is used as a basis for black dye, and it has also some medicinal value.

214. Garcinia morella.—It is supposed that Siam gamboge is obtained from this tree, also that known as Ceylon gamboge. The juice is collected by incising the stems, or by breaking young twigs of the tree and securing the yellow gum resinous exudations in hollow bamboos, where it is allowed to harden. It is employed by artists in water colors and as a varnish for lacquer work.

215. Garcinia pictoria.—A fatty matter known as gamboge butter is procured from the seeds of this tree in Mysore. They are pounded in a stone mortar, then boiled till the butter or oil rises to the surface. It is used as a lamp oil, and sometimes in food.

216. Gardenia florida and Gardenia radicans.—Cape Jasmines, so called from a supposition that they were natives of the Cape of Good Hope. The genus belongs to the cinchona family. G. *lucida* furnishes a fragrant resin somewhat similar to myrrh. The fruit of G. *campanulata* is used as a cathartic, and also to wash out stains in silks. G. *gummifera* yields a resin something like Elemi.

217. Gastrolobium bilobum.—A leguminous plant, having poisonous properties. In western Australia, where it is a native, farmers often lose their cattle through their eating the foliage. Cats and dogs that eat the flesh of these poisoned cattle are also poisoned. G. *obtusum* and G. *spinosum* possess similar properties.

218. Genipa americana.—This belongs to the cinchona family, and produces the fruit called genipap or marmalade box. It is about the size of an orange, and has an agreeable flavor. The juice of the fruit yields a bluish-black dye, called Canito or Lana-dye. This color is very permanent, and is much used by Indians in South America.

219. Geonoma schottiana.—A pretty Brazilian palm; the leaves are used for thatching huts, and other parts of the plant are utilized.

220. Gouania domingensis.—A plant of the buckthorn family, known in Jamaica as Chaw-Stick, on account of its thin branches being chewed as an agreeable stomachic. Tooth brushes are made by cutting pieces of the stem to convenient lengths and fraying out the ends. A tooth powder is prepared by pulverizing the dried stems. It is said to possess febrifugal properties, and owing to its pleasant bitter taste it is used for flavoring cooling beverages.

221. Grevillea robusta.—The silk oak tree of Australia; a tree that attains a large size, and is remarkable for the graceful beauty of its foliage.

222. Grewia asiatica.—This Indian tree represents a genus of plants of considerable economic value. This particular species yields a profusion of small red fruits which are used for flavoring drinks, having a pleasant acid flavor. The fibrous inner bark is employed by the natives for making fishing nets, ropes, twine, and for other similar purposes.

223. Grias cauliflora.—The anchovy pear of Jamaica. The fruit is pickled and eaten like the mango, having a similar taste.

224. Guaiacum officinale.—The wood of this tree is called Lignum Vitæ. A resin, called gum guaiacum, exudes from the stem, and is otherwise obtained from the wood by artificial means. It is of a greenish-brown color, [24] with a balsamic fragrance, and is remarkable for the changes of color it undergoes when brought into contact with various substances. Gluten gives it a blue tint: nitric acid and chlorin change it successively to green, blue, and brown. The resin is used medicinally as also are the bark and wood.

225. Guazuma tomentosa.—This plant is nearly allied to the chocolate-nut tree, and yields fruits that abound in mucilage, as also does the bark of the young shoots. The mucilage is given out in water, and has been used as a substitute for gelatin or albumen in clarifying cane juice in the manufacture of sugar. The timber is light, and is employed for the staves of sugar hogsheads; it is known in Jamaica as bastard cedar. A strong fiber is obtained from the young shoots.

226. Guilielma speciosa.—The peach palm of Venezuela. The fruits are borne in large drooping bunches, and their fleshy outer portion contains starchy matter, which forms a portion of the food of the natives. They are cooked and eaten with salt, and are said to resemble a potato in flavor. A beverage is prepared by fermenting them in water, and the meal obtained from them is made into bread. The wood of the old trees is black, and so hard as to turn the edge of an ax.

227. Hæmatoxylon campechianum.—The logwood tree. This dyestuff is largely used by calico printers and other dyeing manufacturers. It is also used as an ingredient in some writing inks. The heart wood is the part used for dyeing. This is cut into chips which yield their color to water and alcohol. The colors are various according to treatment, giving violet, yellow, purple, and blue, but the consumption of logwood is for black colors, which are obtained by alum and iron bases.

228. Hardenbergia monophylla.—An Australian climbing plant of the leguminous family. The long, carrot-shaped, woody root was called, by the early settlers in that country, sarsaparilla, and is still used in infusion as a substitute for that root.

229. Hartighsea spectabilis.—A New Zealand tree, called Wahahe by the natives, who employ the leaves as a substitute for hops, and also prepare from them a spirituous infusion as a stomachic medicine.

230. Heliconia bihai.—A plant of the order *Musaceæ*, from South America. The young shoots are eaten by the natives, and the fruits are also collected and used as food. It also furnishes a useful fiber.

231. Hevea brasiliensis.—A tree of tropical America growing in damp forests, especially in the Amazon valley, which, together with other trees called siphonia furnish the Para rubber, or American caoutchouc. The sap is collected from incisions made in the tree during the dry season, and is poured over clay molds and dried by gentle heat, successive pourings being made till a sufficiently thick layer is produced.

232. Hibiscus rosa sinensis.—The flowers of this malvaceous plant contain a quantity of astringent juice, and, when bruised, rapidly turn black or deep purple; they are used by the Chinese ladies for dyeing their hair and eyebrows, and in Java for blacking shoes.

233. Hibiscus sabdariffa.—This species is known in the West Indies as red sorrel, on account of the calyxes and capsules having an acid taste. They are made into cooling drinks, by sweetening and fermentation. The bark contains a strong useful fiber which makes good ropes if not too much twisted. It is also known as the Roselle plant.

234. Hibiscus tiliaceus.—A plant common to many tropical countries. Its wood is extremely light when dry, and is employed by the Polynesians for getting fire by friction, which is said to be a very tedious and tiresome operation, and difficult to accomplish. Good fiber is also obtained from the bark.

235. Hippomane mancinella.—This is the poisonous manchineel tree of South America and other tropical regions. The virulent nature of the juice of this tree has given it a reputation equal to that forced upon the upas tree of Java. The juice is certainly very acrid, and even its smoke, when burning, causes temporary blindness. The fruit is equally dangerous, and from its beautiful appearance is sometimes partaken of by those who are unaware of its deleterious properties, but its burning effects on the lips soon causes them to desist. Indians are said to poison their arrows with the juice of this tree.

236. Hura crepitans.—This tropical plant is known as the sandbox tree. Its deep-furrowed, rounded, hard-shelled fruit is about the size of an orange, and when ripe and dry, it bursts open with a sharp noise like the report of [25] a pistol; hence, it is also called the monkey's dinner bell. An emetic oil is extracted from the seeds, and a venomous, milky juice is abundant in all parts of the plant.

237. Hymenæa courbaril.—The locust tree of the West Indies; also called algarroba in tropical regions. This is one of the very largest growing trees known, and living trees in Brazil are supposed to have been growing at the commencement of the Christian era. The timber is very hard, and is much used for building purposes. A valuable resin, resembling the anime of Africa, exudes from the trunk, and large lumps of it are found about the roots of old trees.

238. Hyphæne thebaica.—The doum, or doom palm, or gingerbread of Egypt; it grows also in Nubia, Abyssinia, and Arabia. The fibrous, mealy husks of the seeds are eaten, and taste almost like gingerbread. In the Thebias this palm forms extensive forests, the roots spreading over the lurid ruins of one of the largest and most splendid cities of the ancient world.

239. Icica heptaphylla.—The incense tree of Guiana, a tall-growing tree, furnishing wood of great durability. It is called cedar wood on account of its fragrant odor. The balsam from the trunk is

highly odoriferous, and used in perfumery, and is known as balsam of acouchi; it is used in medicine. The balsam and branches are burned as incense in churches.

240. Ilex paraguayensis.—This is the tea plant of South America, where it occupies the same important position in the domestic economy of the country as the Chinese tea does in this. The *maté* is prepared by drying and roasting the leaves, which are then reduced to a powder and made into packages. When used, a small portion of the powder is placed in a vessel, sugar is added, and boiling water poured over the whole. It has an agreeable, slightly aromatic odor, rather bitter to the taste, but very refreshing and invigorating to the human frame after severe fatigue. It acts in some degree as an aperient and diuretic, and in overdoses produces intoxication. It contains the same active principle, theine as tea and coffee, but not their volatile and empyreumatic oils.

241. Illicium anisatum.—This magnoliaceous plant is a native of China, and its fruit furnishes the star anise of commerce. In China, Japan, and India it is used as a condiment in the preparation of food, and is chewed to promote digestion, and the native physicians prescribe it as a carminative. It is the flavoring ingredient of the preparation *Anisette de Bordeaux*. Its flavor and odor are due to a volatile oil, which is extracted by distillation, and sold as oil of anise, which is really a different article.

242. Illicium floridanum.—A native of the Southern States. The leaves are said to be poisonous; hence, the plant is sometimes called poison bag. The bark has been used as a substitute for cascarilla.

243. Illicium religiosum.—A Japanese species, which reaches the size of a small tree, and is held sacred by the Japanese, who form wreaths of it with which to decorate the tombs of their deceased friends, and they also burn the fragrant bark as incense. Their watchmen use the powdered bark for burning in graduated tubes, in order to mark the time, as it consumes slowly and uniformly. The leaves are said to possess poisonous properties.

244. Indigofera tinctoria.—The indigo plant, a native of Asia, but cultivated and naturalized in many countries. The use of indigo as a dye is of great antiquity. Both Dioscorides and Pliny mention it, and it is supposed to have been employed by the ancient Egyptians. The

indigo of commerce is prepared by throwing the fresh cut plants into water, where they are steeped for twelve hours, when the water is run off into a vessel and agitated in order to promote the formation of the blue coloring matter, which does not exist ready formed in the tissues of the plant, but is the result of the oxidation of other substances contained in them. The coloring matter then settles at the bottom; it is then boiled to a certain consistency and afterwards spread out on cloth frames, where it is further drained of water and pressed into cubes or cakes for market.

245. Ipomœa purga.—A species of jalap is obtained from this convolvulaceous plant; this is a resinous matter contained in the juices.

246. Iriartella setigera.—A South American palm growing in the underwood of the forests on the Amazon and Rio Negro. The Indians use its slender stems for making their blow pipes or gravatanas, through which they blow small poisoned arrows with accuracy to a considerable distance. [26]

247. Jambosa malaccensis.—This Indian plant belongs to the myrtle family. It produces a good-sized edible fruit known as the Malay apple.

248. Jasminum sambac trifoliatum.—A native of South America. The flowers are very fragrant, and an essential oil, much used in perfumery under the name of jasmine oil, is obtained from this and other species.

249. Jatropha clauca.—An East Indian plant the seeds of which when crushed furnish an oil which is used in medicine.

250. Jatropha curcas.—The physic nut tree of tropical America. This plant contains a milky, acrid, glutinous juice, which forms a permanent stain when dropped on linen, and which might form a good marking ink. Burning oil is expressed from the seeds in the Philippine Islands; the oil, boiled with oxide of iron, is used in China as a varnish. It is used in medicine in various ways, the leaves for fomentations, the juice in treating ulcers, and the seeds as purgatives.

251. Jubæa spectabilis.—The coquito palm of Chili. The seed or nut is called cokernut, and has a pleasant, nutty taste. These are

used by the Chilian confectioners in the preparation of sweetmeats, and by the boys as marbles, being in shape and size like them. The leaves are used for thatching, and the trunks or stems are hollowed out and converted into water pipes. A sirup called Miel de Palma or palm honey, is prepared by boiling the sap of this tree to the consistency of treacle, and is much esteemed for domestic use as sugar. The sap is obtained by cutting off the crown of leaves when it immediately begins to flow and continues for several months provided a thin slice is shaved off the top every morning. Full-grown trees will thus yield 90 gallons.

252. Kæmpferia galanga.—This plant belongs to the family of gingers. The root stocks have an aromatic fragrance and are used medicinally in India as well as in the preparation of perfumery. The flowers appear before the leaves upon very short stems.

253. Kigelia pinnata.—This plant is interesting from the circumstance of its being held sacred in Nubia, where the inhabitants celebrate their religious festivals under it by moonlight, and poles made of its wood are erected as symbols of special veneration before the houses of their great chiefs. The fruits, which are very large, when cut in half and slightly roasted, are employed as an outward application to relieve pains.

254. Krameria triandra.—This is one of the species that yield the rhatany roots of commerce. In Peru an extract is made from this species, which is a mild, easily assimilated, astringent medicine. It acts as a tonic, and is used in intermittent and putrid fevers. It is also styptic, and when applied in plasters is used in curing ulcers. The color of the infusion of the roots is blood-red, on which account it is used to adulterate, or rather it forms an ingredient in the fabrication of port wine.

255. Kydia calycina.—An Indian plant of the family *Byttneriaceæ*. The bark is employed in infusion as a sudorific and in cutaneous diseases, and its fibrous tissue is manufactured into cordage.

256. Lagetta lintearia.—The lace-bark tree of Jamaica. The inner bark consists of numerous concentric layers of fibers, which interlace in all directions, and thus present a great resemblance to lace. Articles of apparel are made of it. Caps, ruffles, and even complete suits of lace are made with it. It bears washing with common soap,

and when bleached in the sun acquires a degree of whiteness equal to the best artificial lace. Ropes made of it are very durable and strong.

257. Lansium domesticum.—A low-growing tree of the East Indies, which is cultivated to some extent for its fruit, which is known in Java and Malacca as lanseh fruit, and is much esteemed for its delicate aroma; the pulp is of somewhat firm consistence and contains a cooling, refreshing juice.

258. Lapageria rosea.—A twining plant from Chili. The flowers are very beautiful, and are succeeded by berries, which are said to be sweet and eatable. The root has qualities closely resembling sarsaparilla and used for the same purpose.

259. Latania rubra.—A very beautiful palm from the Mauritius. The fruit contains a small quantity of pulp, which is eaten by the natives, but is not considered very palatable by travelers.

260. Lawsonia inermis.—This is the celebrated henna of the East. The use of the powdered leaves as a cosmetic is very general in Asia and northern Africa, [27] the practice having descended from very remote ages, as is proved by the Egyptian mummies, the parts dyed being usually the finger and toe nails, the tips of the fingers, the palms of the hands, and soles of the feet, receiving a reddish color, considered by Oriental belles as highly ornamental. Henna is prepared by reducing the leaves to powder, and when used is made into a pasty mass with water and spread on the part to be dyed, being allowed to remain for twelve hours. The plant is known in the West Indies as Jamaica Mignonette.

261. Lecythis ollaria.—This tree produces the hard urn-shaped fruits known in Brazil as monkey cups. The seeds are eatable and sold as Sapucaia nuts. The fruit vessels are very peculiar, being 6 inches in diameter and having closely fitting lids, which separate when the seeds are mature. The bark is composed of a great number of layers, not thicker than writing paper, which the Indians separate and employ as cigar wrappers.

262. Leptospermum lanigerum.—A plant known throughout Australia as Captain Cook's tea tree, from the circumstance that, on the first landing of this navigator in that country, he employed a

decoction of the leaves of this plant as a corrective to the effects of scurvy among his crew, and this proved an efficient medicine. Thickets of this plant, along the swampy margin of streams, are known as Tea-tree scrubs. It is also known among the natives as the Manuka plant. The wood is hard and heavy, and was formerly used for making sharp-pointed spears. It belongs to the myrtle family of plants.

263. Licuala acutifida.—This palm is a native of the island of Pulo-Penango, and yields canes known by the curious name of Penang Lawyers. It is a low-growing plant, its stems averaging an inch in diameter. The stems are converted into walking canes by scraping their rough exteriors and straightening them by means of fire heat.

264. Limonia acidissima.—An East India shrub which produces round fruits about the size of damson plums, of a yellowish color, with reddish or purplish tints. They are extremely acid, and the pulp is employed in Java as a substitute for soap.

265. Livistona australis.—This is one of the few palms found in Australia. The unexpanded leaves, prepared by being scalded and dried in the shade, are used for making hats, while the still younger and more tender leaves are eaten like cabbage.

266. Lucuma mammosum.—This sapotaceous plant is cultivated for its fruit, which is called marmalade, on account of its containing a thick agreeably flavored pulp, bearing some resemblance in appearance and taste to quince marmalade. A native of South America.

267. Maba geminata.—The ebony wood of Queensland. The heart wood is black, and the outside wood of a bright red color. It is close-grained, hard, heavy, elastic and tough, and takes a high polish.

268. Macadamia ternifolia.—An Australian tree which produces an edible nut called the Queensland nut. This fruit is about the size of a walnut, and within a thick pericarp, a smooth brown-colored nut, inclosing a kernel of a rich and agreeable flavor, resembling in some degree that of a filbert.

269. Machærium firmum.—A South American tree which furnishes a portion of the rosewood of commerce. Various species of

the genus, under the common Brazilian name of Jaccaranda, are said to yield this wood, but there is some uncertainty about the origin of the various commercial rosewoods.

270. Maclura tinctoria.—The fustic tree. Large quantities of the bright yellow wood of this tree are exported from South America for the use of dyers, who obtain from it shades of yellow, brown, olive, and green. A concentrated decoction of the wood deposits, on cooling, a yellow crystalline matter called Morine. This tree is sometimes called old fustic, in order to distinguish it from another commercial dye called young fustic, which is obtained in Europe from a species of Rhus.

271. Macropiper methysticum.—A plant of the pepper family, which furnishes the root called Ava by the Polynesians. It has narcotic properties, and is employed medicinally, but is chiefly remarkable for the value attached to it as a narcotic and stimulant beverage, of which the natives partake before they commence any important business or religious rites. It is used by chewing the root and extracting the juice, and has a calming rather than an intoxicating effect. It is a filthy preparation, and only partaken of by the lower classes of Feejeeans. [28]

272. Macrozamia denisonii.—An Australian cycad, the seeds of which contain a large amount of farina, or starchy matter, which formerly supplied a considerable amount of food for the natives of that country. The fresh seeds are very acrid, but when steeped in water and roasted they become palatable and nutritious.

273. Malpighia glabra.—A low-growing tree of the West Indies, which produces an edible fruit called the Barbadoes cherry.

274. Mammea Americana.—The fruit of this tree, under the name of mammee apple, is very much esteemed in tropical countries. It often attains a size of 6 or 8 inches in diameter and is of a yellow color. The outer rind and the pulp which immediately surrounds the seeds are very bitter, but the intermediate is sweet and aromatic. The seeds are used as anthelmintics, an aromatic liquor is distilled from the flowers, and the acrid, resinous gum distilled from the bark is used to destroy insects.

275. Manettia cordifolia.—This climbing-plant is a native of South America, and belongs to the family of *Cinchonaceæ*. The rind of the root has emetic properties, and is used in Brazil for dropsy and other diseases. It is also exported under the name of Ipecacuan, chiefly from Buenos Ayres.

276. Mangifera indica.—The mango, in some of its varieties esteemed as the most delicious of tropical fruits, while many varieties produce fruit whose texture resembles cotton and tastes of turpentine. The unripe fruit is pickled. The pulp contains gallic and citric acid. The seeds possess anthelmintic properties. A soft gum resin exudes from the wounded bark, which is used medicinally.

277. Manicaria saccifera.—Bussu palm of South America. Its large leaves are used for thatching roofs, for which purpose they are well fitted and very durable. The fibrous spathe furnishes a material of much value to the natives. This fibrous matter when taken off entire is at once converted into capital bags, in which the Indian keeps the red paint for his toilet, or the silk cotton for his arrows, or he stretches out the larger ones to make himself a cap of nature's own weaving, without seam or joint.

278. Manihot utilissima.—This euphorbiaceous plant yields cassava or mandiocca meal. It is extensively cultivated in tropical climates and supplies a great amount of food. The root is the part used, and in its natural condition is a most virulent poison, but by grating the roots to a pulp the poison is expelled by pressure, and altogether dissipated by cooking. The expressed juice, when allowed to settle, deposits the starch known as tapioca.

279. Maranta arundinacea.—The arrowroot plant, cultivated for its starch. The tubers being reduced to pulp with water, the fecula subsides, and is washed and dried for commerce. It is a very pure kind of starch, and very nutritious. The term arrowroot is said to be derived from the fact that the natives of the West Indies use the roots of the plant as an application to wounds made by poison arrows.

280. Mauritia flexuosa.—The Moriche, or Ita palm, very abundant on the banks of the Amazon, Rio Negro, and Orinoco Rivers. In the delta of the latter it occupies swampy tracts of ground, which are at times completely inundated, and present the appearance of forests

rising out of the water. These swamps are frequented by a tribe of Indians called Guaranes, who subsist almost entirely upon the produce of this palm, and during the period of the inundations suspend their dwellings from the tops of its tall stems. The outer skin of the young leaves is made into string and cord for the manufacture of hammocks. The fermented sap yields palm wine, and another beverage is prepared from the young fruits, while the soft inner bark of the stem yields a farinaceous substance like sago.

281. Maximiliana regia.—An Amazonian palm called Inaja. The spathes are so hard that, when filled with water, they will stand the fire, and are sometimes used by the Indians as cooking utensils. The Indians who prepare the kind of rubber called bottle rubber, make use of the hard stones of the fruit as fuel for smoking and drying the successive layers of milky juice as it is applied to the mold upon which the bottles are formed. The outer husk, also, yields a kind of saline flour used for seasoning their food.

282. Melaleuca minor.—A native of Australia and the islands of the Indian Ocean. The leaves, being fermented, are distilled, and yield an oil known as cajuput or cajeput oil, which is green, and has a strong aromatic odor. It is [29] valuable as an antispasmodic and stimulant, and at one time had a great reputation as a cure for cholera. In China the leaves are used as a tonic in the form of decoction.

283. Melicocca bijuga.—This sapindaceous tree is plentiful in tropical America and the West Indies, and is known as the Genip tree. It produces numerous green egg-shaped fruits, an inch in length, possessing an agreeable vinous and somewhat aromatic flavor, called honey berries or bullace plums. The wood of the tree is hard and heavy.

284. Melocactus communis.—Commonly called the Turk's Cap cactus, from the flowering portion on the top of the plant being of a cylindrical form and red color, like a fez cap. Notwithstanding that they grow in the most dry sterile places, they contain a considerable quantity of moisture, which is well known to mules, who resort to them when very thirsty, first removing the prickles with their feet.

285. Mesembryanthemum crystallinum.—The ice plant, so called in consequence of every part of the plant being covered with small watery pustules, which glisten in the sun like fragments of ice.

Large quantities of this plant are collected in the Canaries and burned, the ashes being sent to Spain for the use of glass makers. *M. edule* is called the Hottentot's fig, its fruit being about the size of a small fig, and having a pleasant, acid taste when ripe. *M. tortuosum* possesses narcotic properties, and is chewed by the Hottentots to induce intoxication. The fruits possess hygrometric properties, the dried, shriveled, capsules swelling out and opening so as to allow of the escape of the seeds when moistened by rain, which at the same time fits the soil for their germination.

286. Mikania guaco.—A composite plant which has gained some notoriety as the supposed Cundurango, the cancer-curing bark. It has long been supposed to supply a powerful antidote for the bite of venomous serpents.

287. Mimusops balata.—The Bully tree. This sapotaceous plant attains a great size in Guiana and affords a dense, close-grained, valuable timber. Its small fruits, about the size of coffee berries, are delicious when ripe. The flowers also yield a perfume when distilled in water, and oil is expressed from the seeds.

288. Mimusops elengi.—A native of Ceylon, where its hard, heavy, durable timber is used for building purposes. The seed also affords a great amount of oil.

289. Monodora grandiflora.—An African plant belonging to the Anonaceæ. It produces large fruit, which contains a large quantity of seeds about the size of the Scarlet-Runner bean. They are aromatic and impart to the fruit the odor and flavor of nutmeg; hence they are also known as calabash nutmegs.

290. Monstera deliciosa.—This is a native of southern Mexico and yields a delicious fruit with luscious pineapple flavor. The outer skin of the fruit, if eaten, causes a stinging sensation in the mouth. This is easily removed when the fruit is ripe. The leaves are singularly perforated with holes at irregular intervals, from natural causes not sufficiently explained. In Trinidad the plant is called the Ceriman.

291. Moringa pterygosperma.—A native of the East Indies, where it bears the name of horse-radish tree. The seeds are called ben nuts

and supply a fluid oil, highly prized by watchmakers, called oil of ben. The root is pungent and stimulant and tastes like horse-radish.

292. Moronobea coccinea.—The hog gum tree, which attains the height of 100 feet. A fluid juice exudes from incisions in the trunk and hardens into a yellow resin. It is said the hogs in Jamaica when wounded rub the injured part against the tree so as to cover it with the gum, which possesses vulnerary properties; hence its name. The resin has been employed as a substitute for copaiba balsam, and plasters are made of it.

293. Mucuna pruriens.—A tall climbing plant of the West Indies and other warm climates. It is called the cowage, or cow-itch, on account of the seed pods being covered with short brittle hairs, the points of which are finely serrated, causing an unbearable itching when applied to the skin, which is relieved by rubbing the part with oil. It is employed as a vermifuge. In East Africa it is called Kitedzi. The sea beans found on the coast of Florida are the seeds of *Mucuna altissima*. In Cuba these are called bulls' eyes. [30]

294. Murraya exotica.—A Chinese plant of the orange family. The fruit is succulent, and the white flowers are very fragrant. They are used in perfumery.

295. Musa cavendishii.—This is a valuable dwarf species of the banana from southern China. It bears a large truss of fine fruit, and is cultivated to some extent in Florida, where it endures more cold than the West India species and fruits more abundantly.

296. Musa ensete.—This Abyssinian species forms large foliage of striking beauty. The food is dry and uneatable; but the base of the flower stalk is eaten by the natives.

297. Musa sapientum.—The banana plant. This has been cultivated and used as food in tropical countries from very remote times, and furnishes enormous quantities of nutritious food, and serves as a staple support to a large number of the human race. The expressed juice is in some countries made into a fermented liquor and the young shoots eaten as a vegetable.

298. Musa textilis.—This furnishes the fiber known as manilla hemp, and is cultivated in the Philippine Islands for this product. The finer kinds of the fiber are woven into beautiful shawls and the

coarser manufactured into cordage for ships. The fiber is obtained from the leaf-stalks.

299. Mussænda frondosa.—This cinchonaceous plant is a native of Ceylon. The bark and leaves are esteemed as tonic and febrifuges in the Mauritius, where they are known as wild cinchona. The leaves and flowers are also used as expectorants, and the juice of the fruit and leaves is used as an eyewash.

300. Myristica moschata.—The nutmeg tree. The seed of this plant is the nutmeg of commerce, and mace is the seed cover of the same. When the nuts are gathered they are dried and the outer shell of the seed removed. The mace is also dried in the sun and assumes a golden yellow color. The most esteemed nutmegs come from Penang. At one time the nutmeg culture was monopolized by the Dutch, who were in the habit of burning them when the crop was too abundant, in order to keep up high prices.

301. Myrospermum peruiferum.—This plant yields the drug known as balsam of Peru, which is procured by making incisions in the bark, into which cotton rags are thrust; a fire is then made round the tree to liquefy the balsam. The balsam is collected by boiling the saturated rags in water. It is a thick, treacly looking liquid, with fragrant aromatic smell and taste, and is not used so much in medicine as it formerly was.

302. Myrospermum toluiferum.—A South American tree, also called Myroxylon, which yields the resinous drug called balsam of Tolu. This substance is fragrant, having a warm, sweetish taste, and burns with an agreeable odor. It is used in perfumery and in the manufacture of pastilles, also for flavoring confectionery, as in Tolu lozenges.

303. Myrtus communis—The common myrtle. This plant is supposed to be a native of western Asia, but now grows abundantly in Italy, Spain, and the south of France. Among the ancients the myrtle was held sacred to Venus and was a plant of considerable importance, wreaths of it being worn by the victors of the Olympic games and other honored personages. Various parts of the plant were used in medicine, in cookery, and by the Tuscans in the preparation of myrtle wine, called *myrtidanum*. It is still used in perfumery, and a highly perfumed distillation is made from the flowers.

The fruits are very aromatic and sweet, and are eaten fresh or dried and used as a condiment.

304. Nandina domestica.—A shrub belonging to the family of berberries. It is a native of China and Japan, where it is extensively cultivated for its fruits. It is there known as Nandin.

305. Nauclea gambir.—A native of the Malayan Islands, which yields the Gambir, or Terra Japonica of commerce. This is prepared by boiling the leaves in water until the decoction thickens, when it is poured into molds, where it remains until it acquires the consistency of clay; it is then cut into cubes and thoroughly dried. It is used as a masticatory in combination with the areca nut and betel leaf, and also for tanning purposes.

306. Nectandra leucantha.—The greenheart, or bibiru tree of British Guiana, furnishing bibiru bark, which is used medicinally as a tonic and febrifuge, its properties being due to the presence of an uncrystallizable alkaloid, also found in the seeds. The seeds are also remarkable for containing [31] upwards of 50 per cent of starch, which is made into a kind of bread by the natives. The timber of this tree is extensively employed in shipbuilding, its great strength and durability rendering it peculiarly well suited for this purpose.

307. Nepenthes distillatoria.—This pitcher plant is a native of Ceylon. The pitchers are partly filled with water before they open; hence it was supposed to be produced by some distilling process. In Ceylon the old, tough, flexible stems are used as willows.

308. Nephelium litchi.—This sapindaceous tree produces one of the valued indigenous fruits of China. There are several varieties; the fruit is round, about an inch and a half in diameter, with a reddish-colored, thin, brittle shell. When fresh they are filled with a sweet, white, transparent, jelly-like pulp. The Chinese are very fond of these fruits and consume large quantities of them, both in the fresh state and when dried and preserved.

309. Nerium oleander.—This is a well-known plant, often seen in cultivation, and seemingly a favorite with many. It belongs to a poisonous family and is a dangerous poison. A decoction of its leaves forms a wash, employed in the south of Europe to destroy vermin; and its powdered wood and bark constitute the basis of an

efficacious rat-poison. Children have died from eating the flowers. A party of soldiers in Spain, having meat to roast in camp, procured spits and skewers of the tree, which there attains a large size. The wood having been stripped of its bark, and brought in contact with the meat, was productive of fatal consequences, for seven men died out of the twelve who partook of the meat and the other five were for some time dangerously ill.

310. Notelæa ligustrina.—The Tasmanian iron wood tree. It is of medium growth and furnishes wood that is extremely hard and dense, and used for making sheaves for ships' blocks, and for other articles that require to be of great strength. The plant belongs to the olive family.

311. Ochroma Lagopus.—A tree that grows about 40 feet high, along the seashores in the West Indies and Central America, and known as the cork wood. The wood is soft, spongy, and exceedingly light, and is used as a substitute for cork, both in stopping bottles and as floats for fishing nets. It is also known as Balsa.

312. Œnocarpus batava.—A South American palm, which yields a colorless, sweet-tasted oil, used in Para for adulterating olive oil, being nearly as good for this purpose as peanut oil, so largely used in Europe. A palatable but slightly aperient beverage is prepared by triturating the fruits in water, and adding sugar and mandiocca flour.

313. Olea europæa.—The European olive, which is popularly supposed to furnish *all* the olive oil of commerce. It is a plant of slow growth and of as slow decay. It is considered probable that trees at present existing in the Vale of Gethsemane are those which existed at the commencement of the Christian era. The oil is derived from the flesh of the fruit, and is pressed out of the bruised pulp; inferior kinds are from second and third pressings. The best salad oil is from Leghorn, and is sent in flasks surrounded by rush-work. Gallipoli oil is transported in casks, and Lucca in jars. The pickling olives are the unripe fruits deprived of a portion of their bitterness by soaking in water in which lime and wood ashes are sometimes added, and then bottled in salt and water with aromatics.

314. Ophiocaryon paradoxum.—The snake nut tree of Guiana, so called on account of the curious form of the embryo of the seed,

which is spirally twisted, so as to closely resemble a coiled-up blacksnake. The fruits are as large as those of the black walnut, and although they are not known to possess any medical properties, their singular snake-like form has induced the Indians to employ them as an antidote to the poison of venomous snakes. The plant belongs to the order of *Sapindaceæ*.

315. Ophiorrhiza mungos.—A plant belonging to the cinchona family, the roots of which are reputed to cure snake bites. They are intensely bitter, and from this circumstance they are called earth-galls by the Malays.

316. Ophioxylon serpentinum.—A native of the East Indies, where the roots are used in medicine as a febrifuge and alexipharmic.

317. Opuntia cochinellifera.—A native of Mexico, where it is largely cultivated in what are called the Nopal plantations for the breeding of the cochineal insect. This plant and others are also grown for a similar purpose in the [32] Canary Islands and Madeira. Some of these plantations contain fifty thousand plants. Cochineal forms the finest carmine scarlet dye, and at least there are 2,000 tons of it produced yearly, in value worth $2,000 per ton.

318. Opuntia tuna.—This plant is a native of Mexico and South America generally. It reaches a height of 15 to 20 feet and bears reddish-colored flowers, followed by pear-shaped fleshy fruits 2 or 3 inches long, and of a rich carmine color when ripe. It is cultivated for rearing the cochineal insect. The fruits are sweet and juicy; sugar has been made from them. The juice is used as a water-color and for coloring confectionery.

319. Oreodaphne Californica.—The mountain laurel, or spice bush, of California. When bruised it emits a strong, spicy odor, and the Spanish Americans use the leaves as a condiment.

320. Oreodoxa oleracea.—The West Indian cabbage palm, which sometimes attains the height of 170 feet, with a straight cylindrical trunk. The semicylindrical portions of the leaf-stalk are formed into cradles for children, or made into splints for fractures. Their inside skin, peeled off while green, and dried, looks like vellum, and can be written upon. The heart of young leaves, or cabbage, is boiled as

a vegetable or pickled, and the pith affords sago. Oil is obtained from the fruit.

321. Ormosia dasycarpa.—This is the West Indian bead tree, or necklace tree, the seeds of which are roundish, beautifully polished, and of a bright scarlet color, with a black spot at one end resembling beads, for which they are substitutes, being made into necklaces, bracelets, or mounted in silver for studs and buttons. It is a leguminous plant.

322. Osmanthus fragrans.—This plant has long been cultivated as *Olea fragrans*. The flowers have a fine fragrance, and are used by the Chinese to perfume tea. It appears that they consider the leaves also valuable, for they are frequently found in what is expected to be genuine tea.

323. Pachira alba.—A South American tree the inner bark of which furnishes a strong useful fiber, employed in the manufacture of ropes and various kinds of cordage. The petals of the flowers are covered with a soft silky down which is used for stuffing cushions and pillows.

324. Pandanus utilis.—The screw pine of the Mauritius, where it is largely cultivated for its leaves, which are manufactured into bags or sacks for the exportation of sugar. They are also used for making other domestic vessels and for tying purposes.

325. Pappea capensis.—A small tree of the soapberry or sapindaceous family, a native of the Cape of Good Hope, where the fruit is known as the wild plum, from the pulp of which a vinous beverage and excellent vinegar are prepared, and an eatable, though slightly purgative, oil is extracted from the seeds. The oil is also strongly recommended for baldness and scalp affections.

326. Papyrus antiquorum.—The paper-reed of Asia, which yielded the substances used as paper by the ancient Egyptians. The underground root-stocks spread horizontally under the muddy soil, continuing to throw up stems as they creep along. The paper was made from thin slices, cut vertically from the apex to the base of the stem, between its surface and center. The slices were placed side by side, according to the size required, and then, after being wetted

and beaten with a wooden instrument until smooth, were pressed and dried in the sun.

327. Paritium elatum.—The mountain mahoe, a malvaceous plant, that furnishes the beautiful lace-like bark called Cuba bast, imported by nurserymen for tying their plants. It was at one time only seen as employed in tying together bundles of genuine Havana cigars. It forms a tree 40 feet or more in height, and yields a greenish-blue timber, highly prized by cabinet-makers.

328. Parkia africana.—The African locust tree, producing seeds which the natives of Soudan roast, and then bruise and allow to ferment in water until they become putrid, when they are carefully washed, pounded into powder, and made into cakes, which are said to be excellent, though having a very unpleasant smell. The pulp surrounding the seeds is made into a sweet farinaceous preparation. [33]

329. Parkinsonia aculeata.—This leguminous plant is called Jerusalem Thorn. Although a native of Southern Texas and Mexico, it is found in many tropical countries, and is frequently used for making hedges. Indians in Mexico employ it as a febrifuge and sudorific and also as a remedy for epilepsy.

330. Parmentiera cereifera.—In the Isthmus of Panama this plant is termed the Candle tree, because its fruits, often 4 feet long, look like yellow candles suspended from the branches. They have a peculiar, apple-like smell, and cattle that partake of the leaves or fruit have the smell communicated to the beef if killed immediately.

331. Passiflora quadrangularis.—The fruit of this plant is the Granadilla of the tropics. The pulp has an agreeable though rather mawkish taste. The root is said to possess narcotic properties, and is used in the Mauritius as an emetic.

332. Paullinia sorbilis.—The seeds of this climbing sapindaceous plant furnish the famous guarana of the Amazon and its principal tributaries. The ripe seeds, when thoroughly dried, are pounded into a fine powder, which made into dough with water, is formed into cylindrical rolls, from 5 to 8 inches long, becoming very hard when dry. It is used as a beverage, which is prepared by grating about half a teaspoonful of one of the cakes into about a teacup of

water. It is much used by Brazilian miners, and is considered a preventive of all manner of diseases. It is also used by travelers, who supply themselves with it previous to undertaking lengthy or fatiguing journeys. Its active principle is identical with theine, of which it contains a larger quantity than exists in any other known plant, being more than double that contained in the best black tea.

333. Pavetta borbonica.—This belongs to the quinine family. The roots are bitter, and are employed as a purgative; the leaves are also used medicinally.

334. Pedilanthus tithymaloides.—This euphorbiaceous plant has an acrid, milky, bitter juice; the root is emetic, and the dried branches are used medicinally.

335. Pereskia aculeata.—The Barbadoes gooseberry, which belongs to the family *Cactaceæ*. It grows about 15 feet in height, and produces yellow-colored, eatable, and pleasant-tasted fruit, which is used in the West Indies for making preserves.

336. Persea gratissima.—The avocado or alligator pear, a common tree in the West Indies. The fruits are pear-shaped, covered with a brownish-green or purple skin. They are highly esteemed where grown, but strangers do not relish them. They contain a large quantity of firm pulp, possessing a buttery or marrow-like taste, and are frequently called vegetable marrow. They are usually eaten with spice, lime-juice, pepper, and salt. An abundance of oil, for burning and for soap-making, may be obtained from the pulp. The seeds yield a deep, indelible black juice, which is used for marking linen.

337. Phœnix dactylifera.—The date palm, very extensively grown for its fruit, which affords the principal food for a large portion of the inhabitants of Africa, Asia, and southern Europe, and likewise of the various domestic animals—dogs, horses, and camels being alike partial to it. The tree attains to a great age, and bears annually for two hundred years. The huts of the poorer classes are constructed of the leaves: the fiber surrounding the bases of their stalks is used for making ropes and coarse cloth; the stalks are used for the manufacture of baskets, brooms, crates, walking sticks, etc., and the wood for building substantial houses; the heart of young leaves is eaten as a vegetable; the sap affords an intoxicating beverage. It may be further mentioned that the date was, probably, the palm

which supplied the "branches of palm trees" mentioned by St. John (xii, 13) as having been carried by the people who went to meet Christ on his triumphal entry into Jerusalem, and from which Palm Sunday takes its name.

338. Phormium tenax.—This plant is called New Zealand flax, on account of the leaves containing a large quantity of strong, useful fiber, which is used by the natives of that country for making strings, ropes, and articles of clothing. The plant could be grown in this climate, and would no doubt be largely cultivated if some efficient mode of separating the fiber could be discovered. [34]

339. Photinia japonica.—The Japanese Medlar, or Chinese Loquat. It bears a small oval fruit of an orange color when ripe, having a pleasant subacid flavor. It stands ordinary winters in this climate, and forms a fine evergreen, medium-sized tree.

340. Physostigma venenosum.—A strong leguminous plant, the seeds of which are highly poisonous, and are employed by the natives of Old Calabar as an ordeal. Persons suspected of witchcraft or other crimes are compelled to eat them until they vomit or die, the former being regarded as proof of innocence, and the latter of guilt. Recently the seeds have been found to act powerfully in diseases of the eye.

341. Phytelephas macrocarpa.—The vegetable ivory plant, a native of the northern parts of South America. The fruit consists of a collection of six or seven drupes; each contains from six to nine seeds, the vegetable ivory of commerce. The seeds at first contain a clear, insipid liquid; afterwards it becomes milky and sweet, and changes by degrees until it becomes hard as ivory. Animals eat the fruit in its young green state; a sweet oily pulp incloses the seeds, and is collected and sold in the markets under the name of Pipa de Jagua. Vegetable ivory may be distinguished from animal ivory by means of sulphuric acid, which gives a bright red color with the vegetable ivory, but none with the animal ivory.

342. Picrasma excelsa.—This yields the bitter wood known as Jamaica Quassia. The tree is common in Jamaica, where it attains the height of 50 feet. The wood is of a whitish or yellow color, and has an intensely bitter taste. Although it is used as a medicine in cases of weak digestion, it acts as a narcotic poison on some animals, and the

tincture is used as fly poison. Cups made of this wood, when filled with water and allowed to remain for some time, will impart tonic properties to the water.

343. Pinckneya pubens. — This cinchonaceous plant is a native of the Southern States and has a reputation as an antiperiodic. It is stated that incomplete examinations have detected *cinchonine* in the bark. It has been used successfully as a substitute for quinine. A thorough examination of this plant seems desirable so that its exact medical value may be ascertained.

344. Piper betel. — This plant belongs to the *Piperaceæ*. Immense quantities of the leaves of this plant are chewed by the Malays. It tinges the saliva a bright red and acts as a powerful stimulant to the digestive organs and salivary glands; when swallowed it causes giddiness and other unpleasant symptoms in persons unaccustomed to its use.

345. Piper nigrum. — This twining shrub yields the pepper of commerce. It is cultivated in the East and West Indies, Java, etc., the Malabar being held in the highest esteem. The fruit when ripe is of a red color, but it is gathered before being fully ripe and dried in the sun, when it becomes black and shriveled. White pepper is the same fruit with the skin removed. When analyzed, pepper is found to contain a hot acrid resin and a volatile oil, as well as a crystalline substance called *piperin*, which has been recommended as a substitute for quinine.

346. Pistacia lentiscus. — The mastic tree, a native of southern Europe, northern Africa, and western Asia. Mastic is the resin of the tree and is obtained by making transverse incisions in the bark, from which it exudes in drops and hardens into small semitransparent tears. It is consumed in large quantities by the Turks for chewing to strengthen the gums and sweeten the breath. It is also used for varnishing.

347. Pistacia terebinthus. — The Cyprus turpentine tree. The turpentine flows from incisions made in the trunk and soon becomes thick and tenacious, and ultimately hardens. Galls gathered from this tree are used for tanning purposes, one of the varieties of morocco leather being tanned with them.

348. Pistacia vera.—The pistacia tree, which yields the eatable pistachio nuts. It is a native of western Asia. The nuts are greatly eaten by the Turks and Greeks, as well as in the south of Europe, either simply dried like almonds or made into articles of confectionery.

349. Pithecolobium saman.—This leguminous plant yields eatable pods, which are fed to cattle in Brazil. Some Mexican species produce pods that are boiled and eaten, and certain portions contain saponaceous properties. The pods are sometimes called Manila tamarinds. The leaves of this tree fold closely up at night, so that they do not prevent the radiation of heat from the surface of the ground, and dew is therefore deposited underneath its branches. The grass on the surface of the ground underneath this tree being thus wet [35] with dew, while that under other trees is found to be dry, has given it the name of rain tree, under the supposition that the leaves dropped water during the night.

350. Pittosporum undulatum.—A plant from New Zealand, which reaches a considerable size, and furnishes a wood similar to boxwood. The flowers are very fragrant.

351. Plagianthus betulinus.—The inner bark of the young branches of this plant yields a very fine fiber, sometimes called New Zealand cotton, though more like flax than cotton; it is the Akaroa of the New Zealanders. In Tasmania it bears the name of Currajong. Good cordage and twine for fishing nets are made from this fiber. A superior paper pulp is prepared from the wood; it is also employed in making handles to baskets, rims for sieves, and hoops for barrels.

352. Platonia insignis.—A Brazilian tree which bears a fruit known in that country as Pacoury-uva. The pulp of this fruit is semiacid, very delicious, and is employed in making preserves. The seeds embedded in this pulp have the flavor of almonds.

353. Plumbago scandens.—The root of this plant is called Herbe du Diable in San Domingo; it is acrid in the highest degree, and is a most energetic blistering agent when fresh.

354. Plumeria alba.—A South American plant. The flowers are used in perfumery, and furnish the scent known as Frangipane or Frangipani. In Jamaica the plant is known as red jasmine.

355. Pogostemon patchouly.—This plant affords the celebrated patchouli perfume. The peculiar odor of patchouli is disagreeable to some, but is very popular with many persons. The odoriferous part of the plant is the leaves and young tops, which yield a volatile oil by distillation, from which an essence is prepared; satchels of patchouli are made of coarsely powdered leaves. Genuine Indian shawls and Indian ink were formerly distinguished by their odor of this perfume, but the test does not now hold good. Ill effects, such as loss of sleep, nervous attacks, etc., have been ascribed to its extensive use.

356. Pongamia glabra.—Some years ago this tree was recommended as suitable for avenue-planting in the south of France. In India an oil called poonga is expressed from the seeds, which is much used for mixing with lamp oil. It is of a deep yellow color, and is fluid at temperatures above 60° F., but below that it becomes solid.

357. Portlandia grandiflora.—This plant belongs to the cinchonaceous family, and is said to possess properties similar to those of the true cinchona. The bark is exceedingly bitter.

358. Psidium cattleyanum.—This is the purple guava from China. The fruits are filled with juicy, pale flesh, of a very agreeable acid-sweet flavor.

359. Psidium pyriferum.—The West Indian guava, a well-known fruit in the tropics, but only known here in the shape of guava jelly. The wood of the tree has a fine, close grain, and has been experimented with as a substitute for boxwood for engraving purposes, but it is too soft to stand the pressure of printing.

360. Psychotria leucantha.—A plant belonging to the cinchona family. Emetic properties are assigned to the roots, which are also used in dyeing. Native of Peru.

361. Pterocarpus marsupium.—This tree affords gum-kino, which is obtained by making incisions in the bark, from which the juice exudes and hardens into a brittle mass, easily broken into small angular, shining fragments of a bright ruby color. It is highly astringent. The wood is hard and valuable for manufacturing purposes.

362. Punica granatum. — The pomegranate, a native of northern Africa and western Asia. The fruit is valued in warm countries on account of its delicious cooling and refreshing pulp. Numerous varieties are grown, some being sweet and vinous, and others acid or of a bitter, stringent taste; the color also varies from light to dark red. The bark of the root abounds in a peculiar principle called *punicin*. This bark appears to have been known to the ancients, and used by them as a vermifuge, and is still used in Hindostan as a specific against tapeworm. The rind of the fruit of the bitter varieties contains a large amount of tannin, and is used for tanning morocco leather. The flowers yield a red dye. [36]

363. Quassia amara. — The wood of this plant furnishes Surinam quassia. It is destitute of smell, but has an intensely bitter taste, and is used as a tonic. The root has also reputed medicinal value, as also have the flowers.

364. Quillaja saponaria. — The Quillai or Cully of the Chilians. Its bark is called soap-bark, and is rough and dark-colored externally, but internally consists of numerous regular whitish or yellowish layers, and contains a large quantity of carbonate of lime and other mineral matters. It is also rich in *saponine*, and is used for washing clothes; 2 ounces of the bark is sufficient to wash a dress. It also removes all spots or stains, and imparts a fine luster to wool; when powdered and rubbed between the hands in water, it makes a foam like soap. It is to be found in commerce.

365. Randia aculeata. — A small tree native of the West Indies, also found in southern Florida. In the West Indies the fruit is used for producing a blue dye, and medicinal properties are assigned to the bark.

366. Raphia tædigera. — The Jupati palm. The leaf-stalks of this plant are used by the natives of the Amazon for a variety of purposes, such as constructing inside walls, making boxes and baskets, etc. *R. vinifera*, the Bamboo palm, is similarly used by the Africans, who also make a very pliable cloth of the undeveloped leaves. Palm wine is one of the products of the genus.

367. Ravenala madagascariensis. — This plant is called the Traveler's tree, probably on account of the water which is stored up in the large cup-like sheaths of the leaf-stalks, and which is sought for by

travelers to allay their thirst. The broad leaves are used in Madagascar as thatch to cover their houses. The seeds are edible, and the blue, pulpy aril surrounding them yields an essential oil.

368. Rhapis flabelliformis.—The ground rattan palm. This is supposed to yield the walking-canes known as rattan, which is doubted. It is a native of southern China, and is also found in Japan, where it is known by the name of Kwanwortsik.

369. Rhizophora mangle.—This plant is known as the mangrove, possibly because no man can live in the swampy groves that are covered with it in tropical countries. The seeds germinate, or form roots before they quit the parent tree, and drop into the mud as young trees. The old plants send out aerial roots into the water, upon which the mollusca adhere, and as the tide recedes they are seen clinging to the shoots, verifying the statements of old travelers that they had seen oysters growing on trees. All parts of this tree contain tannin. The bark yields dyes, and in the West Indies the leaves are used for poulticing wounds. The fruit is edible; a coarse, brittle salt is extracted from the roots, and in the Philippines the bark is used as a febrifuge.

370. Rottlera tinctoria.—This plant belongs to the order *Euphorbiaceæ*, and reaches the size of a small tree in the Indian Archipelago and southern Australia. From the surface of the trilobed capsules of this plant, which are about the size of peas, a red, mealy powder is obtained, well known in India as kamala, and which is used by Hindoo silk-dyers, who obtain from it a deep, bright, durable orange or flame color of great beauty. This is obtained by boiling the powder in a solution of carbonate of soda. When the capsules are ripe the red powder is brushed off and collected for sale, no other preparation being necessary to preserve it. It is also used medicinally as an anthelmintic and has been successfully used in cases of tapeworm. A solution removes freckles and pustules and eruptions on the skin.

371. Ruellia indigotica.—This small bush is extensively cultivated in China for the preparation of a blue coloring-matter of the nature of indigo. The pigment is prepared from the entire plant by a process similar to that employed in procuring the common indigo. It is sold in China in a pasty state. The water in which the plant is

steeped is mixed with lime and rapidly agitated, when the coloring deposits at the bottom of the vessel.

372. Sabal adansonii.—This dwarf palm is a native of the Southern States. The leaves are made into fans, and the soft interior of the stem is edible.

373. Sabal umbraculifera.—This is a West Indian palm; the leaves are used for various purposes, such as making mats, huts, etc.

374. Saccharum officinarum.—The sugar cane. Where the sugar cane was first cultivated is unknown, but it is supposed to have been in the East Indies, for the Venetians imported it from thence by the Red Sea prior to the year 1148. It is supposed to have been introduced into the islands of Sicily, Crete, [37] Rhodes, and Cyprus by the Saracens, as abundance of sugar was made in these islands previous to the discovery of the West Indies in 1492 by the Spaniards, and the East Indies and Brazil by the Portuguese in 1497 and 1560. It was cultivated afterwards in Spain, in Valentia, Granada, and Murcia by the Moors. In the fifteenth century it was introduced into the Canary Islands by the Spaniards and to Madeira by the Portuguese, and thence to the West India Islands and to Brazil. The Dutch began to make sugar in the island of St. Thomas in the year 1610 and in Jamaica in 1644. Its culture has since become general in warm climates and its use universal.

375. Saguerus saccharifer.—The arenga palm, which is of great value to the Malays. The black horsehair like fiber surrounding its leaf-stalks is made into cordage; a large amount of toddy or palm wine is obtained by cutting off the flower spikes, which, when inspissated, affords sugar, and when fermented a capital vinegar. Considerable quantities of inferior sago and several other products of minor importance are derived from this palm.

376. Sagus rumphii.—This palm produces the sago of commerce, which is prepared from the soft inner portion of the trunk. It is obtained by cutting the trunk into small pieces, which are split and the soft substance scooped out and pounded in water till the starchy substance separates and settles. This is sago meal; but before being exported it is made into what is termed pearl sago. This is a Chinese process, principally carried on at Singapore. The meal is washed, strained, and spread out to dry; it is then broken up, pounded, and

sifted until it is of a regular size. Small quantities being then placed in bags, these are shaken about until it becomes granulated or pearled.

377. Salvadora persica.—This is supposed to be the plant that produced the mustard seed spoken of in the Scriptures.

378. Sandoricum indicum.—A tropical tree, sometimes called the Indian sandal tree, which produces a fruit like an apple, of agreeable acid flavor. The root of the tree has some medicinal value.

379. Sanseviera guineensis.—Called the African bowstring hemp, from the fibers of the leaves being used for bowstrings.

380. Santalum album.—This tree yields the true sandalwood of India. This fragrant wood is in two colors, procured from the same tree; the yellow-colored wood is from the heart and the white-colored from the exterior, the latter not so fragrant. The Chinese manufacture it into musical instruments, small cabinets, boxes, and similar articles, which are insect proof. From shavings of the wood an essential oil is distilled, which is used in perfumery.

381. Sapindus saponaria.—The soapberry tree. The fruit of this plant is about the size of a large gooseberry, the outer covering or shell of which contains a saponaceous principle in sufficient abundance to produce a lather with water and is used as a substitute for soap. The seeds are hard, black, and round, and are used for making rosaries and necklaces, and at one time were covered for buttons. Oil is also extracted from the seeds and is known as soap oil.

382. Sapium indicum.—A widely distributed Asiatic tree which yields an acrid, milky juice, which, as also the leaves of the plant, furnishes a kind of dye. The fruit in its green state is acid, and is eaten as a condiment in Borneo.

383. Sapota achras.—The fruit of this plant is known in the West Indies as the sapodilla plum. It is highly esteemed by the inhabitants; the bark of the tree is astringent and febrifugal; the seeds are aperient and diuretic.

384. Sapota mulleri.—The bully or balata tree of British Guiana, which furnishes a gum somewhat intermediate between India rubber and gutta-percha, being nearly as elastic as the first without the

brittleness and friability of the latter, and requiring a high temperature to melt or soften it.

385. Schinus molle. — The root of this plant is used medicinally and the resin that exudes from the tree is employed to astringe the gums. The leaves are so filled with resinous fluid that when they are immersed in water it is expelled with such violence as to have the appearance of spontaneous motion in consequence of the recoil. The fruits are of the size of pepper corns and are warm to the taste. The pulp surrounding the seeds is made into a kind of beverage by the Mexican Indians. The plant is sometimes called Mexican pepper. [38]

386. Schotia speciosa. — A small tree of South Africa called Boerboom at the Cape of Good Hope. The seeds or beans are cooked and eaten as food. The bark is used for tanning purposes and as an astringent in medicine.

387. Seaforthia elegans. — This palm is a native of the northern part of Australia, where it is utilized by the natives. The seeds have a granular fibrous rind, and are spotted and marked like a nutmeg.

388. Selaginella lepidophylla. — This species of club moss is found in southern California, and has remarkable hygrometric qualities. Its natural growth is in circular roseate form, and fully expanded when the air is moist, but rolling up like a ball when it becomes dry. It remains green and acts in this peculiar manner for a long time after being gathered. Of late years numbers have been distributed throughout the country under the names of "Rose of Jericho" and "Resurrection Plant." This is, however, quite distinct from the true Rose of Jericho, *Anastatica hierochuntica*, a native of the Mediterranean region, from Syria to Algeria. This plant, when growing and in flower, has branches spread rigidly, but when the seed ripens the leaves wither, and the whole plant becomes dry, each little branch curling inward until the plant appears like a small ball; it soon becomes loosened from the soil, and is carried by the winds over the dry plains, and is often blown into the sea, where it at once expands. It retains this property of expanding when moistened for at least ten years.

389. Semecarpus anacardium. — The marking nut tree of India. The thick, fleshy receptacle bearing the fruit is of a yellow color

when ripe, and is roasted and eaten. The unripe fruit is employed in making a kind of ink. The hard shell of the fruit is permeated by a corrosive juice, which is used on external bruises and for destroying warts. The juice, when mixed with quick-lime, is used to mark cotton or linen with an indelible mark. When dry it forms a dark varnish, and among other purposes it is employed, mixed with pitch and tar, in the calking of ships. The seeds, called Malacca beans, or marsh nuts, are eaten, and are said to stimulate the mental powers, and especially the memory; and finally they furnish an oil used in painting.

390. Serissa fœtida.—A cinchonaceous shrub, having strong astringent properties. The roots are employed in cases of diarrhea, also in ophthalmia and certain forms of ulcers. It is a native of Japan and China.

391. Shorea robusta.—This tree produces the Saul wood of India, which has a very high reputation, and is extensively employed for all engineering purposes where great strength and toughness are requisite. It is stronger and much heavier than teak. An oil is obtained from the seeds, and a resin similar to Dammar resin is likewise obtained from the tree.

392. Sida pulchella.—A plant of the mallow family; the bark contains fibrous tissues available for the manufacture of cordage. The root of S. acuta is esteemed by the Hindoos as a medicine, and particularly as a remedy for snake bites. The light wood of these species is used to make rocket sticks.

393. Simaba cedron.—A native of New Grenada, where it attains the size of a small tree, and bears a large fruit containing one seed; this seed, which looks like a blanched almond, is known in commerce as the cedron. As a remedy for snake bites it has been known from time immemorial in New Grenada. It is mentioned in the books of the seventeenth century. Recently it has obtained a reputation as a febrifuge, but its value as an antidote to the bites of snakes and scorpions is universally believed, and the inhabitants carry a seed with them in all their journeyings; if they happen to be bitten by any venomous reptile they scrape about two grains of the seed in brandy or water and apply it to the wound, at the same time taking a like dose internally. This neutralizes the most dangerous poisons.

394. Simaruba officinalis.—This tree yields the drug known as Simaruba bark, which is, strictly speaking, the rind of the root. It is a bitter tonic. It is known in the West Indies as the mountain damson.

395. Siphonia elastica.—The South American rubber plant, from which a great portion of the caoutchouc of commerce is obtained. There are several species of siphonia which, equally with the above, furnish the India rubber exported from Para. The caoutchouc exists in the tree in the form of a thin, white milk, which exudes from incisions made in the trunk, and is poured over molds, which were formerly shaped like jars, bottles, or [39] shoes, hence often called bottle rubber. As it dries, the coatings of milky juice are repeated until the required thickness is obtained, and the clay mold removed. It belongs to the extensive family *Euphorbiaceæ*.

396. Smilax medica.—This plant yields *Mexican* sarsaparilla, so called to distinguish it from the many other kinds of this drug. The plant is a climber, similar to the smilax of our woods.

397. Spondias mombin.—This yields an eatable fruit called hog plum in the West Indies. The taste is said to be peculiar, and not very agreeable to strangers. It is chiefly used to fatten swine. The fruit is laxative, the leaves astringent, and the seeds possess poisonous qualities. The flower buds are used as a sweetmeat with sugar.

398. Strelitzia reginæ.—A plant of the Musa or banana family. The flowers are very beautiful for the genus. It is a native of the Cape of Good Hope. The seeds are gathered and eaten by the Kaffirs.

399. Strychnos nux-vomica.—This is a native of the Coromandel coast and Cochin-China. It bears an orange-like fruit, containing seeds that have an intensely bitter taste, owing to the presence of two most energetic poisons, *strychnine* and *brucine*. The pulp surrounding the seeds is said to be harmless, and greedily eaten by birds. The wood of the plant is hard and bitter, and possesses similar properties to the seeds, but in a less degree. It is used in India in intermittent fevers and in cases of snake bites. S. *tiente* is a Java shrub, the juice of which is used in poisoning arrows. S. *toxifera* yields a frightful poison called Ourari or Wourari, employed by the natives of Guiana. This is considered to be the most potent sedative in nature. Several species of *Strychnos* are considered infallible rem-

edies for snake bites; hence are known as snakewood. *S. pseudoquina*, a native of Brazil, yields Colpache bark, which is much used in that country in cases of fever, and is considered equal to quinine in value. It does not contain strychnine, and its fruits are edible. *S. potatorum* furnishes seeds known in India as clearing-nuts, on account of their use in clearing muddy water. St. Ignatius beans are supposed to be yielded by a species of Strychnos, from the quantity of strychnine contained in the seeds.

400. Swietenia mahagoni.—This South American plant furnishes the timber known in commerce as mahogany. The bark is considered a febrifuge, and the seeds prepared with oil were used by the ancient Aztecs as a cosmetic. The timber is well known, and much used in the manufacture of furniture.

401. Tacca pinnatifida.—This is sometimes called South Sea arrowroot. The tubers contain a great amount of starch, which is obtained by rasping them and macerating four or five days in water, when the fecula separates in the same manner as sago. It is largely used as an article of diet throughout the tropics, and is a favorite ingredient for puddings and cakes.

402. Tamarindus indica.—The tamarind tree. There are two varieties of this species. The East Indian variety has long pods, with six to twelve seeds. The variety cultivated in the West Indies has shorter pods, containing one to four seeds. Tamarinds owe their grateful acidity to the presence of citric, tartaric, and other vegetable acids. The pulp mixed with salt is used for a liniment by the Creoles of the Mauritius. Every part of the plant has had medicinal virtues ascribed to it. Fish pickled with tamarinds are considered a great delicacy. It is said that the acid moisture exhaled by the leaves injures the cloth of tents that remain under them for any length of time. It is also considered unsafe to sleep under the trees.

403. Tanghinia venenifera.—This plant is a native of Madagascar, and of the family *Apocynaceæ*. Formerly, when the custom of trial by ordeal was more prevalent than now, the seeds of this plant were in great repute, and unlimited confidence was placed in the poisonous seeds as a detector of guilt. The seeds were pounded, and a small piece swallowed by each person to be tried; those in whom it caused vomiting were allowed to escape, but when it was retained

in the stomach, it would quickly prove fatal, and their guilt was thus held to be proven.

404. Tasmannia aromatica.—The bark of this plant possesses aromatic qualities, closely resembling Winter's bark. The small black fruits are used as a substitute for pepper. [40]

405. Tectona grandis.—The teak tree. Teak wood has been extensively employed for shipbuilding in the construction of merchant vessels and ships of war; its great strength and durability, the facility with which it can be worked, and its freedom from injury by fungi, rendering it peculiarly suitable for these purposes. It is a native of the East India Islands, and belongs to the order *Verbenaceæ*.

406. Terminalia catappa.—The astringent fruits of this tropical plant are employed for tanning and dyeing, and are sometimes met with in commerce under the name of myrobalans, and used by calico printers for the production of a permanent black. The seeds are like almonds in shape and whiteness, but, although palatable, have a peculiar flavor.

407. Tetranthera laurifolia.—This plant is widely dispersed over tropical Asia and the islands of the Eastern Archipelago. Its leaves and young branches abound in a viscid juice, and in Cochin-China the natives bruise and macerate them until this becomes glutinous, when it is used for mixing with plaster, to thicken and render it more adhesive and durable. Its fruits yield a solid fat, used for making candles, although it has a most disagreeable odor.

408. Thea viridis.—This is the China tea plant, whose native country is undetermined. All kinds and grades of the teas of commerce are made from this species, although probably it has some varieties. Black and green teas are the result of different modes of preparation; very much of the green, however, is artificially colored to suit the foreign trade. The finest teas do not reach this country; they will not bear a sea voyage, and are used only by the wealthy classes in China and Russia. The active principles of the leaves are theine and a volatile oil, to which latter the flavor and odor are due. So far as climate is concerned for the existence of the tea plant in the United States, it will stand in the open air without injury from Virginia southwards. A zero frost will not kill it. But with regard to its production as a profitable crop, the rainfall in no portion of the States is

sufficient to warrant any attempt to cultivate the plant for commercial purposes. But this does not prevent its culture as a domestic article, and many hundreds of families thus prepare all the tea they require, from plants it may be from the pleasure ground or lawn, where the plant forms one of the best ornaments.

409. Theobroma cacao.—This plant produces the well-known cacao, or chocolate, and is very extensively cultivated in South America and the West India Islands. The fruit, which is about 8 to 10 inches in length by 3 to 5 in breadth, contains between fifty and a hundred seeds, and from these the cacao is prepared. As an article of food it contains a large amount of nutritive matter, about 50 per cent being fat. It contains a peculiar principle, which is called *theobromine.*

410. Theophrasta jussiæi.—A native of St. Domingo, where it is sometimes called Le petit Coca. The fruit is succulent, and bread is made from the seeds.

411. Thespesia populnea.—A tropical tree, belonging to the mallow family. The inner bark of the young branches yields a tough fiber, fit for cordage, and used in Demerara for making coffee bags, and the finer pieces of it for cigar envelopes. The wood is considered almost indestructible under water, and its hardness and durability render it valuable for various purposes. The flower buds and unripe fruits yield a viscid yellow juice, useful as a dye, and a thick, deep, red-colored oil is expressed from the seeds.

412. Thevetia neriifolia.—This shrubby plant is common in the West Indies and in many parts of Central America. Its bark abounds in a poisonous milky juice, and is said to possess powerful properties. A clear, bright, yellow-colored oil, called Exile oil, is obtained, by expression, from the seeds.

413. Thrinax argentea.—This beautiful palm is called the Silver Thatch palm of Jamaica, and is said to yield the leaves so extensively used in the manufacture of hats, baskets, and other articles. It is also a native of Panama, where it is called the broom palm, its leaves being there made into brooms.

414. Tillandsia zebrina.—A South American plant of the pineapple family; the bottle-like cavity at the base of the leaves will some-

times contain a pint or more of water, and has frequently furnished a grateful drink to thirsty travelers. [41]

415. Tinospora cordifolia.—A climbing plant, so tenacious of life that when the stem is cut across or broken, a rootlet is speedily sent down from above, which continues to grow until it reaches the ground. A bitter principle, *calumbine*, pervades the plant. An extract called galuncha is prepared from it, considered to be a specific for the bites of poisonous insects and for ulcers. The young shoots are used as emetics.

416. Triphasia trifoliata.—A Chinese shrub, with fruit about the size of hazelnuts, red-skinned, and of an agreeable sweet taste; when green, they have a strong flavor of turpentine, and the pulp is very sticky. They are also preserved whole in sirup, and are sometimes called limeberries.

417. Tristania neriifolia.—A myrtaceous plant from Australia, called the turpentine tree, owing to its furnishing a fluid resembling that product.

418. Urceola elastica.—A plant belonging to the *Apocynaceæ*, a native of the islands of Borneo and Sumatra, where its milky juice, collected by making incisions in its soft, thick, rugged bark, or by cutting the trunk into junks, forms one of the kinds of caoutchouc called juitawan, but it is inferior to the South American, chiefly owing to want of care in its preparation, the milky juice being simply coagulated by mixing with salt water, instead of being gradually inspissated in layers on a mold. The fruit contains a pulp which is much eaten by the natives.

419. Urena lobata.—A malvaceous plant, possessing mucilaginous properties, for which it is used medicinally. The bark affords an abundance of fiber, resembling jute rather than flax or hemp.

420. Uvaria odoratissima.—An Indian plant which is supposed to yield the essential oil called Ylang-Ylang, or Alan-gilan. This oil is obtained by distillation from the flowers, and is highly esteemed by perfumers, having an exquisite odor partaking of the jasmine and lilac.

421. Vangueria edulis.—A cinchonaceous plant, the fruits of which are eaten in Madagascar under the name of Voa-vanga. The leaves are used in medicine.

422. Vanilla planifolia.—The vanilla plant, which belongs to the orchid family. The fruit is used by confectioners and others for flavoring creams, liquors, and chocolates. There are several species, but this gives the finest fruit. It is a climbing orchid, and is allowed to climb on trees when cultivated for its fruit. In Mexico, from whence is procured a large portion of the fruit, it is cultivated in certain favorable localities near the Gulf coast, where the climate is warm. Much of the value of the bean depends upon the process of its preparation for the market. In Mexico, where much care is given to this process, the pods are gathered before they are fully ripe and placed in a heap, under protection from the weather, until they begin to shrivel, when they are submitted to a sweating process by wrapping them in blankets inclosed in tight boxes; afterwards they are exposed to the sun. They are then tied into bundles or small bales, which are first wrapped in woolen blankets, then in a coating of banana leaves first sprinkled with water, then placed in an oven heated up to about 140° F. Here they remain for twenty-four to forty-eight hours, according to the size of the pods, the largest requiring the longest time. After this heating they are exposed to the sun daily for fifty or sixty days, until they are thoroughly dried and ready for the market.

423. Vateria indica.—This plant yields a useful gum resin, called Indian copal, piney varnish, white dammar, or gum anine. The resin is procured by cutting a notch in the tree, so that the juice may flow out and become hardened. It is used as a varnish for pictures, carriages, etc. On the Malabar coast it is manufactured into candles, which burn with a clear light and an agreeable fragrance. The Portuguese employ this resin instead of incense. Ornaments are fashioned from it under the name of amber. It is also employed in medicine.

424. Weinmannia racemosa.—A New Zealand tree called Towhia by the natives of that country. Its bark is used for tanning purposes, and as a red and brown dye, which give fast colors upon cotton fabrics.

425. Wrightia tinctoria.—The leaves of this plant furnish an inferior kind of indigo. The wood is beautifully white, close-grained, and ivory-like, and is much used for making Indian toys.

426. Xanthorrhœa arborea.—The grass gum tree of Australia, also called black boy. This is a liliaceous plant, which produces a long flower-stalk, bearing at the top an immense cylindrical flower-spike, and when the short [42] black stem is denuded of leaves, the plants look very like black men holding spears. The leaves afford good fodder for cattle, and the tender white center is used as a vegetable. A fragrant resin, called acaroid resin, is obtained from it.

427. Ximenia americana.—A small tree, found in many warm regions; among others in southern Florida. In Brazil it is called the Native Plum on account of its small yellow fruits, which have a subacid and somewhat astringent aromatic taste. The wood is odoriferous and is used in the West Indies as a substitute for sandalwood.

428. Yucca alœfolia.—The yucca leaves afford a good fiber, and some southern species are known as *bear's grass*. The root stems also furnish a starchy matter, which has been rendered useful in the manufacture of starch.

429. Zamia furfuracea.—This plant belongs to the order *Cycadaceæ*, and is grown to some extent for the starchy matter contained in the stem, which is collected and used as arrowroot; but it is not the true arrowroot, that being produced by a species of *Maranta*.

430. Zamia integrifolia.—The coontie plant of Florida. The large succulent roots afford a quantity of arrowroot, said to be equal to the best of that from Bermuda. The fruit has a coating of an orange-colored pulp, which is said to form a rich edible food. It was from the roots of this plant that the Seminoles of Florida obtained their *white meal*.

431. Zingiber officinale.—This plant is cultivated in most warm countries for the sake of its rhizomes, which furnish the spice called ginger. It is prepared by digging up the roots when a year old, scraping them, and drying them in the sun. Ginger, when broken across, shows a number of little fibers embedded in floury tissue. Its hot pungent taste is due to a volatile oil. It also contains starch and

yellow coloring matter. Ginger is used for various medicinal purposes, and in many ways as a condiment, and in the preparation of cordials and so-called teas.

www.ingramcontent.com/pod-product-compliance
Lightning Source LLC
Chambersburg PA
CBHW030446220526
45464CB00006B/2432